T0135932

Structure-borne Sound Sources in Buildings

Von der Fakultät für Elektrotechnik und Informationstechnik der
Rheinischen-Westfälischen Technischen Hochschule Aachen
zur Erlangung des akademischen Grades eines

DOKTORS DER INGENIEURWISSENSCHAFTEN

genehmigte Dissertation

vorgelegt von

M.Sc.
Matthias Lievens

aus Gent, Belgien

Berichter:

Universitätsprofessor Dr. rer. nat. Michael Vorländer
Universitätsprofessor Dr.-Ing. Dirk Heberling

Tag der mündlichen Prüfung: 3.Mai 2013

Matthias Lievens

Structure-borne Sound Sources in Buildings

Logos Verlag Berlin GmbH

λογος

Aachener Beiträge zur Technischen Akustik

Editor:
Prof. Dr. rer. nat. Michael Vorländer
Institute of Technical Acoustics
RWTH Aachen University
52056 Aachen
www.akustik.rwth-aachen.de

Bibliographic information published by the Deutsche Nationalbibliothek

The Deutsche Nationalbibliothek lists this publication in the Deutsche Nationalbibliografie; detailed bibliographic data are available in the Internet at http://dnb.d-nb.de .

D 82 (Diss. RWTH Aachen University, 2013)

ISBN 978-3-8325-3464-6
ISSN 1866-3052
Vol. 15

Logos Verlag Berlin GmbH
Comeniushof, Gubener Str. 47,
D-10243 Berlin
Tel.: +49 (0)30 / 42 85 10 90
Fax: +49 (0)30 / 42 85 10 92
http://www.logos-verlag.de

Contents

Abstract

Structure-borne sound sources are vibrational sources connected in some way to the building structure. The mechanical excitation of the building structure leads to sound radiation. This is an important source of annoyance in modern light-weight buildings. The prediction of the sound pressure level from structure-borne sound sources is highly complicated because of the complexity involved in the coupling between source and receiver structure. The current standard on characterisation of service equipment in buildings EN 12354-5, can deal with sources on heavy structures (high-mobility source) but to date, there is no engineering method available for the case of coupling between source and receiver.

To develop a practical engineering method it is crucial to determine the important degrees of freedom. The coupling between source and receiver can theoretically be described by a mobility approach through six degrees of freedom at every contact point. This leads to an enormous amount of data that has to be simplified. One of the assumption frequently made in the field of building acoustics is to use the normal components only. This aspect is investigated in this thesis on a case study of a washing machine on a wooden joist floor. The directly measured sound pressure level is compared with the predicted sound pressure level. The prediction is based on normal components only. Moment excitation is minimised by reducing the contact area between source and receiver while the in-plane components are assumed negligible.

In the first part of this thesis, measurements in the coupled state are conducted. It is shown that the normal components are sufficient to predict the sound pressure level. However, this only applies to the coupled state. In the second part, a true prediction is calculated from independently measured source and receiver quantities. The difference between predicted and directly measured sound pressure level leads to considerable errors of up to 20 dB at low frequencies. This shows that the normal components are not sufficient to predict the coupling between a washing machine and a wooden floor.

1 Introduction

The acoustical comfort in buildings can be described by the performance of the building elements in terms of two main criteria: the airborne sound insulation and the structure-borne sound insulation. The latter is subdivided into impact sound and structure-borne sound from service equipment. This distinction originates from the importance. The sound that is heard when people walk over floors in buildings is called impact sound. It is one of the most important aspects for the total acoustical quality of a building and it is usually dealt with as a first priority. The remaining sound that originates from the mechanical excitation of the building by sources like air-conditioning units, water taps, heating systems and pumps is called structure-borne sound from service equipment. A few examples of common structure-borne sound sources in buildings are depicted in Figure 1.1. Airborne sound insulation refers to the performance of building elements (e.g. walls) when it comes to attenuating incoming airborne sound waves.

Structure-borne sound from service equipment has been neglected for many years. With the introduction of higher standards of sound insulation in buildings those secondary structure-borne sound sources have now become important as well. The treatment of those sources is highly complicated due to the fact that structure-borne sound sources interact with the structure they are connected to. This interaction varies depending on the characteristics of the source, the receiver and the point of contact on the receiver structure. In the field of building acoustics there is a need for engineering methods that describe this source/receiver interaction with acceptable accuracy. The methods have to be practical and the expected levels of accuracy have to be well documented.

Those methods are currently not available due to the complexity involved in characterising the sources through measurements and in describing the coupling between a source and the building structure. A lot of research in this field has been done [1, 2, 3, 4, 5] and this thesis is part of it.

Legal requirements

Legislation with respect to acoustical comfort is different all over the world but the descriptors are very strongly related due to the internationally accepted standards for the measurement methods. In Germany, for example, the DIN 4109 [6] stipulates a value of L_{AFmax} of 30dB(A) as requirement for the sound pressure level due to service equipment. The measurement procedure to obtain the value of L_{AFmax} (the maximum A-weighted sound pressure level measured with an integration time-constant of 125 ms) is explained in the standard ISO 10052 [7] or according to a more precise engineering method in ISO 16032 [8]. Different measurement cycles are provided for several sources like water taps, shower cubicles and toilets. For stationary sources the measurement duration is 30 s, impulsive sources are measured during one operation cycle. During this period of time the sound pressure level is measured at three positions in the room and averaged. The requirements in the DIN 4109 only apply to equipment installed in the building. Washing machines, tumble dryers or any vibrational sources that can be removed from the building are not covered by the DIN 4109. However, this does not make those sources less important as potential noise sources.

EN 12354 - Part 5

Measuring L_{AFmax} in the installed condition is usually done to verify whether the legal requirements are met. More important however is the prediction of sound pressure levels before the service equipment is actually installed in a building or even during the process of designing a building. The state-of-the-art prediction method in Europe is the standard EN 12354 [9, 10, 11, 12, 13, 14]. The estimation of the acoustic performance of buildings from the performance of the separate elements is the subject of this standard. Part 1-3 of this standard on airborne-sound, impact sound and outdoor sound are successfully being used by building companies to predict the sound insulation in buildings at the design stage, e.g. in commercial software like BASTIAN [15].

Part 5 of this standard is devoted to the prediction of sound pressure levels due to service equipment and is closely related to the work in this thesis. The EN12354-5 describes how the sound pressure levels can be calculated from the characteristic structure-borne sound power of the source and a coupling factor that depends on the source and the receiver. However the applicability of this part of the standard strongly relies on future research. Currently the measurement method to obtain the characteristic structure-borne sound power only exist for high-mobility sources, i.e. for sources with a high-mobility compared to the mobility of the building structure. The measurement method is defined in EN 15657–1 [16]. The source under test is positioned on a standardised concrete reception plate and the velocity is measured at random locations on the plate during source operation [17, 18]. By combining the measured velocity with the loss factor of the plate (determined through the measured reverberation time of the plate) the input power can be readily obtained.

In case of coupling between the structure-borne sound source and the building (no high-mobility source) there is no method available to measure the characteristic power of the source. This means that the necessary input data for the prediction according to EN 12354-5 is missing for this kind of sources. The second input that is required for a prediction according to EN 12354 Part 5 is the coupling factor. It is calculated out of the measured mobility of the source and the measured mobility of the building structure. The calculation treats all contact points independently but allows for all six degrees of freedom to be taken into account. It is pointed out in the standard that more experience with measurements on a variety of sources will lead to specific measurement procedures in the future. This again means that to obtain the coupling factor there are currently no practical measurement methods available. Research on this topic proposes the use of two reception plates with different mobilities [19]. Difficulties in finding a high-mobility reception plate that still statically supports the source under test are to be expected with this method. More recent research investigates the indirect source characterisation by measuring the source with and without a reception plate [20]. It is however clear that more research is vital to guarantee the applicability of part 5 of EN15657.

Thesis Objectives

Structure-borne sound sources have to be characterised to enable the prediction of the radiated sound from the structures they are connected to. In buildings the sources will typically be washing machines, fans, air-conditioning units, heating systems, pumps, household appliances and all sorts of vibrational sources used or installed in a building. Those sources are either fixed or positioned on a

(a) air-conditioning unit on a wall (b) air-conditioning units on a roof (c) pumps

(d) communal heating power station (e) communal heating power station integrated in building (f) heating system

Figure 1.1: Common structure-borne sound sources in buildings.

wall or floor and sound is radiated from those receiver structures. **It is the objective of this thesis to investigate how well such connected systems can be predicted from independent source and receiver quantities by considering only the normal forces in the direction of the axes through a contact point (translational y-components).** This is considered to be the most precise case in terms of practical source characterisation in building acoustics. It is very unlikely that the effort involved with this precision will ever be made in practice but it is important to know how accurate a prediction can be in that case. A case study of a washing machine on a wooden floor was conducted to answer this question.

The washing machine was chosen to represent a class of vibrational sources being coupled through multiple contact points to a receiver structure in a building. The washing machine as such is not the main subject of investigation. It is likely that the results presented for the washing machine can be applied to a whole class of sources as long as the dimensions of the source and the contact area are similar. This will have to be verified in further case studies.

The characterisation of structure-borne sound sources is not as straightforward as for airborne sound sources. An airborne sound source transmits power through its coupling to the air which is in most practical cases *independent* of the position in a room or the room itself. A structure-borne sound source on the other hand is coupled through one or more connection points to a receiver structure and for every connection point (or different receiver structure) a broad range of mobilities can be encountered. This variety of coupling situations results in a specific power flow for every case. For large differences between the magnitude of the source and receiver mobility two ideal cases can be assumed which simplify the prediction substantially: a high-mobility source and a low-mobility source. However, the mobility of source and receiver are usually in the same order of magnitude if lightweight receiver structures (e.g. timber joist floor, gypsum board walls, interior walls of an aircraft, ...) and their radiation in the audible frequency range are considered. In those cases the interaction between the source and receiver mobility is crucial and a correct prediction of the power flow can only be achieved by combining the source and receiver mobility. A thorough overview on the subject is given by Petersson and Gibbs [21].

A system consisting of a source connected to a receiver structure through multiple contact points can be precisely described with the mobility approach. The approach has been frequently adopted in the past to describe mechanical systems and it is a well-established method in the field of source characterisation [22, 23, 24]. For linear time-invariant systems, it is theoretically possible to account for the source activity through the free velocity or the blocked force. The interaction between the different contact points and the different degrees of freedom can be dealt with by the mobility matrices of the source and the receiver [21, 25]. The exact solution consists of the interaction between 6 N by 6 N source and receiver matrices according to the six degrees of freedom (three rotational and three translational components) at every of the N contact points as shown in Fig. 1.2. The multi-point multi-component method describes the behaviour of the source as a whole through its contact points. For example the moment of the complete source (pitch and roll) is dealt with by a particular phase relation between the translational y-components of the different contact points. In other words, the moments at the contact points (rotational x-components and z-components) are different from the moment of the complete source.

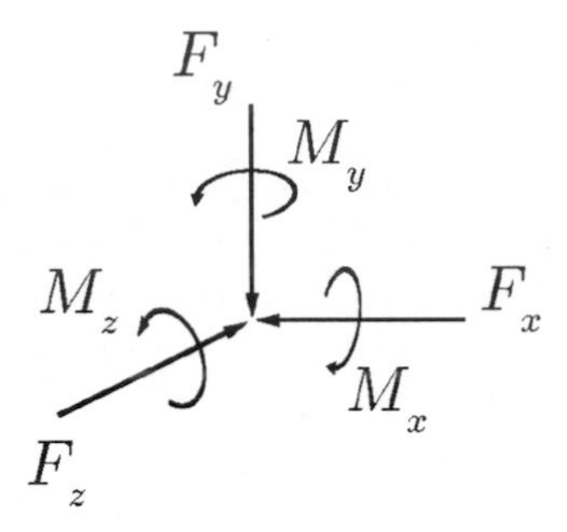

Figure 1.2: The coupling between a source and receiver is defined through six degrees of freedom at every contact point.

By looking at a practical source connected to a plate-like structure (e.g. service equipment connected to walls or floors), it is usually concluded that the in-plane components and the rotational degree of freedom around the axis of the connection can be neglected because of their unlikely contribution to the sound radiation. After this simplification still three degrees of freedom remain important: two moments (rotational x-components and z-components) and the normal components (translational y-components).

The thesis consist of two main parts. In the first part (chapter 3 and 4) it is investigated which degrees of freedom are important for the sound radiation. Because of the complexity involved in measuring all degrees of freedom a method of omission was chosen: only the normal components out of the six degrees of freedom are used in the calculation of the sound pressure. The calculation is then compared to the directly measured pressure (which contains all degrees of freedom by nature) to investigate the influence of the non-normal components on the sound radiation. The method measures all quantities in the coupled state, i.e. the source stays attached to the receiver at all times. To avoid the measurement of the moments and to artificially design case studies in which only the normal components dominate the sources were connected to the floor through a very limited contact area. The contact area was then gradually increased.

The second part in chapter 5 goes one step further and uses only the normal components to predict the coupling (interaction force) between the source and the receiver from independently measured normal source and receiver quantities. This is the ultimate goal of structure-borne sound source characterisation. The vibrational quantities for different coupling situations with different receiver structures are *predicted* from the independent data of the source and the receiver. The predicted interaction is evaluated in the sound field by calculating sound pressure from the interaction forces.

The work in this thesis can be seen as a preliminary study in the context of EN 12354. The case studies are carefully designed to provide ideal test cases that can be used to investigate the influence of the different assumptions by calculation. This allows for the error involved with certain assumptions to be quantified in advance.

The methods in this thesis differ from the approach that uses reception plates (EN 15657-1 [16]). A clear advantage of the reception plate method is the simplicity of the measurement procedure which makes it very suitable for practical engineering methods in the field of building acoustics. The measurements are conducted with accelerometers on a standardised reception plate and the data is evaluated in third octave bands. However, the reception plate method uses average data over several contact points and assumes a certain phase relation between the feet. For fundamental investigations the reception plate should not be used because the intrinsic assumptions of the method introduce errors that cannot be distinguished from any other discrepancies. For this reason all measurements in this thesis are complex field quantities.

2 Theory

This chapter contains material from [26, 27].

The system under investigation consists of a source coupled to a floor. The source excites the floor and the sound is radiated into the receiving room according to Fig. 2.1. At every contact point between the source and the receiver only the normal components out of the prevailing six degrees of freedom are considered throughout the work as shown in Fig. 1.2. However, the theory presented below is written in matrix form and also applies to the general case including six degrees of freedom.

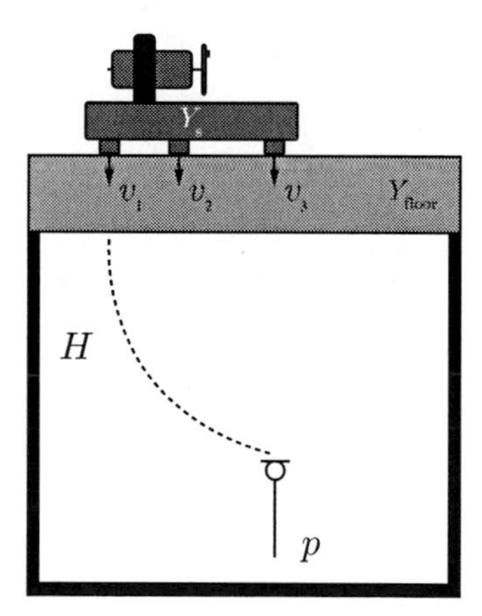

Figure 2.1: Schematic representation of the measurement setup (one degree of freedom is drawn). Those normal components (transversal y-components) are used in the measurements.

The coupling between the source and the receiver is accounted for by the mobility approach [24]. A schematic drawing of the coupled mechanical system is shown in Fig. 2.2. For practical reasons the mobility of the floor is not separated from the room volume. The loading of the volume of air on the floor is always included in $\mathbf{Y}_\mathrm{r}$.

The source activity of a vibrational source can be represented with the free velocity, $\mathbf{v}_\mathrm{f}$, or the blocked force, $\mathbf{F}_\mathrm{b}$. Equivalent circuit diagrams that correspond to those two cases are shown in Fig. 2.2.

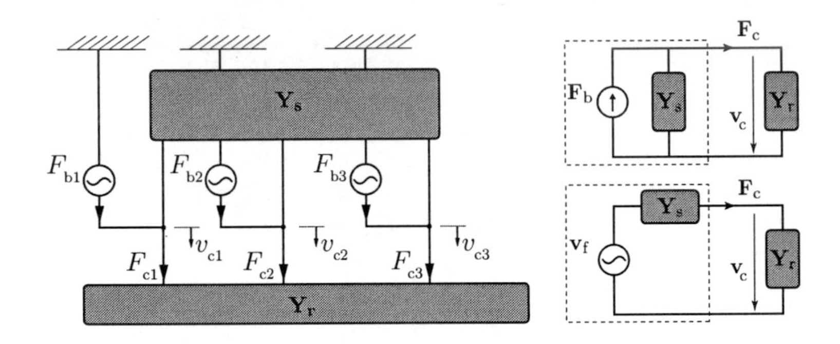

Figure 2.2: Mechanical model (one degree of freedom is drawn) and equivalent circuit diagrams.

The mobility, $\mathbf{Y}$, of a system is the relationship between the velocity and the force at a series of arbitrary terminals according to:

$$\mathbf{v} = \mathbf{Y}\mathbf{F}. \tag{2.1}$$

If a system is excited with force vector, $\mathbf{F}$, it will respond to this force with a certain velocity vector, $\mathbf{v}$. The relation between force and velocity is the mobility, $\mathbf{Y}$, of the system. By applying a force at one contact point at the time the elements of the mobility matrix can be determined. The matrix notation deals with multiple contact points of the system and their interaction and with the interaction between the different degrees of freedom [28]. All variables are complex quantities. A point mobility relates the excitation and response at the same contact point. A transfer mobility is the mobility between two points. The impedance matrix, $\mathbf{Z}$, can be obtained by inverting the mobility matrix.

In the setup under test the source mobility, $\mathbf{Y}_s$ is linked to the receiver mobility, $\mathbf{Y}_r$. The combination of those mobilities equals the coupled mobility, the mobility of the assembled system as shown in Fig. 2.2:

$$\mathbf{Y}_c = (\mathbf{Y}_s{}^{-1} + \mathbf{Y}_r{}^{-1})^{-1}. \tag{2.2}$$

The free velocity of a source is related to its blocked force according to this relationship:

$$\mathbf{v}_f = \mathbf{Y}_s \mathbf{F}_b. \tag{2.3}$$

2.1 Prediction of the interaction force

From inspection of the the equivalent circuit diagrams in Fig. 2.2 the interaction force can be calculated according to

$$\mathbf{F}_c = (\mathbf{Y}_s + \mathbf{Y}_r)^{-1} \mathbf{Y}_s \mathbf{F}_b, \tag{2.4}$$

$$\mathbf{F}_c = (\mathbf{Y}_s + \mathbf{Y}_r)^{-1} \mathbf{v}_f. \tag{2.5}$$

It should be noted that an inversion of a mobility matrix is only valid if all degrees of freedom are taken into account. If, for example, only one degree of freedom is considered the matrix inversion is only valid in the absence of coupling between the degrees of freedom [3].

2.2 Ideal sources

To evaluate the performance of the above predictions in Eq. 2.4 and 2.5 it is interesting to compare them with the case of a high-mobility source and a low-mobility source. Those sources neglect the coupling between the source and the receiver and can be used as an upper limit to the prediction error. For a high-mobility source, $\mathbf{Y}_s \gg \mathbf{Y}_r$ and Eq. 2.4 and 2.5 reduce to

$$\mathbf{F}_c \approx \mathbf{F}_b. \tag{2.6}$$

For a low-mobility source, $\mathbf{Y}_s \ll \mathbf{Y}_r$ and results in

$$\mathbf{F}_c \approx \mathbf{Y}_r{}^{-1} \mathbf{v}_f. \tag{2.7}$$

By use of Eq. 2.3 these ideal cases can both be expressed through $\mathbf{v}_f$ and $\mathbf{F}_b$.

2.3 Transfer paths

A transfer path relates the pressure measured at an arbitrary position to the excitation force of the structure according to:

$$p = \mathbf{HF}. \tag{2.8}$$

The coupling through discrete contact points vanishes in the sound field: the pressure at the observation point is a scalar quantity and the transfer path is a vector. The elements of the transfer path vector have to be obtained one at the time by exciting the contact points consecutively according to

$$H_j = \frac{p}{F_j}, \quad F_k = 0 \quad \text{for}\, k \neq j. \tag{2.9}$$

Classical transfer path analysis makes use of the interaction force, $\mathbf{F}_\mathrm{c}$, to predict the sound pressure. In that case, the transfer paths have to be obtained without the source being installed otherwise the coupling through the source would result in a superimposed force. The fact that interaction forces are used in the calculation implies that a virtual separation of the source and the receiver was carried out. Hence, the frequency response functions (frfs) also have to be measured on the disassembled system.

2.4 Prediction of the sound pressure in the coupled state

2.4.1 Prediction of the sound pressure from the operational velocity: derivation through Cramer's rule

The model shown in Figure 2.3 serves as a simple example of a coupled system that is used to explain the theory. It is not a direct representation of the measurement setup because the source has only two connection points and the observation point 3 is located on the receiver structure and not in the sound field. It is nevertheless closely related as it consists of a structure-borne sound source with mobility Y_s connected to a receiver with mobility Y_r. The source activity is represented by the blocked force.

The coupled system is completely described by the system of equations in (2.10) and the solution is readily obtained by matrix operations. The aim is however

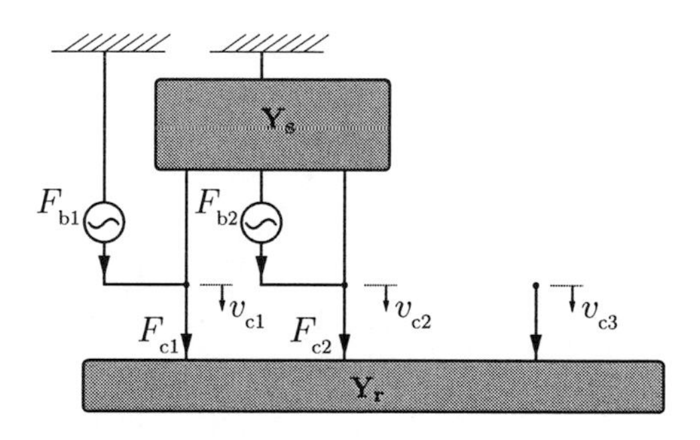

Figure 2.3: Mechanical model of source coupled through two contact points to a receiver.

to calculate the behaviour at an observation point by using measured frequency response functions and *in-situ* measured velocities during operation. In the theoretical example the velocity at the observation point 3 is predicted by using operational velocities v_{c1} and v_{c2} and the relevant frequency response functions between point 1, 2 and 3.

The coupled mobility of the assembled system is a combination of the source and receiver mobility according to Eq. 2.2. In the case of two contact points the system of equations for the normal components of the coupled system is:

$$\mathbf{Y}_{\mathrm{c}}^{-1} \times \begin{bmatrix} v_{\mathrm{c}1} \\ v_{\mathrm{c}2} \\ v_{\mathrm{c}3} \end{bmatrix} = \begin{bmatrix} F_{\mathrm{b}1} \\ F_{\mathrm{b}2} \\ 0 \end{bmatrix}. \tag{2.10}$$

It can be shown that through the use of Cramer's rule [29] the solution of a system of equations can be derived according to:

$$v_j = \frac{1}{|\mathbf{Y}_{\mathrm{c}}^{-1}|} \sum_{k=1}^{n} A_{kj} F_{\mathrm{b}k}. \tag{2.11}$$

A_{kj} denotes the cofactor of the k-th row and the j-th column of matrix $\mathbf{Y}_{\mathrm{c}}^{-1}$. Hence, from the system of equations in (2.10) $v_{\mathrm{c}3}$ can be calculated as a function of v_{c1} and v_{c2} according to Cramer's rule:

$$v_{c3} = \frac{1}{|\mathbf{Y}_c^{-1}|}(A_{13} \cdot F_{b1} + A_{23} \cdot F_{b2}) \tag{2.12}$$

$$\begin{aligned} v_{c1} &= \frac{1}{|\mathbf{Y}_c^{-1}|}(A_{11} \cdot F_{b1} + A_{21} \cdot F_{b2}) \\ v_{c2} &= \frac{1}{|\mathbf{Y}_c^{-1}|}(A_{12} \cdot F_{b1} + A_{22} \cdot F_{b2}) \end{aligned} \tag{2.13}$$

$$v_{c3} = \frac{(A_{13}\,A_{22} - A_{12}\,A_{23})\,v_{c1} \;+\; (A_{11}\,A_{23} - A_{13}\,A_{21})\,v_{c2}}{A_{11}\,A_{22} - A_{12}\,A_{21}} \tag{2.14}$$

The solution shows that the velocity at the third terminal can be calculated by using the cofactors of $\mathbf{Y}_c{}^{-1}$, v_{c1} and v_{c2}.

Cramer's rule is now used again to relate the cofactors to measured frequency response functions Y_c. If the coupled system at rest is excited with F_{bk} at one terminal Cramer's rule states:

$$\frac{v_{cj}}{F_{bk}} = \frac{1}{|\mathbf{Y}_c^{-1}|}A_{kj} = Y_{cjk} \tag{2.15}$$

This means that the kj-cofactor is proportional to a measured frequency response function between terminal k and j. By substituting (2.15) into (2.14) v_3 becomes:

$$v_{c3} = \frac{(Y_{c31}\,Y_{c22} - Y_{c21}\,Y_{c32})\,v_{c1} \;+\; (Y_{c11}\,Y_{c32} - Y_{c31}\,Y_{c12})\,v_{c2}}{Y_{c11}\,Y_{c22} - Y_{c21}\,Y_{c12}} \tag{2.16}$$

Equation (2.16) predicts the velocity at the observation point in terms of the operational velocity v_{c1} and v_{c2} in combination with the frequency response functions Y_c. In practise the operational velocities can easily be measured while the source is running and connected to the receiver. The frequency response functions of the system also have to be measured in its coupled state, i.e. the source is connected to the receiver. This is commonly done with an impulse hammer or a mechanical shaker. It is however important not to change the dynamics of the system with the external exciter.

The above derivation is valid for any structure-borne sound source connected to a receiving structure through one or more terminals. The theory can also easily be extended to n connection points.

Three contact points

For the prediction of the sound pressure as shown in Fig. 2.1, the theory is now to be used on a measurement setup that consists of a structure-borne sound source connected through three feet to a plate. The plate radiates into a receiving room and the observation point is an arbitrary point in the room. The observation point can be seen as an arbitrary point in the linear sytem just like point 3 in the mechanical model. At that point the sound pressure is measured and the frequency response function, H_{ck}, is defined similarly to Equation 2.15 as:

$$H_{ck} = \frac{p}{F_{bk}} \tag{2.17}$$

The resulting pressure due to a source with three excitation terminals is:

$$p = \frac{\begin{pmatrix} Y_{c21}Y_{c32}H_{c3} - Y_{c21}H_{c2}Y_{c33} - Y_{c31}Y_{c22}H_{c3} \\ +Y_{c31}H_{c2}Y_{c23} + H_{c1}Y_{c22}Y_{c33} - H_{c1}Y_{c32}Y_{c23} \end{pmatrix} v_{c1} + \begin{pmatrix} -Y_{c11}Y_{c32}H_{c3} + Y_{c11}H_{c2}Y_{c33} + Y_{c31}Y_{c12}H_{c3} \\ -Y_{c31}H_{c2}Y_{c13} - H_{c1}Y_{c12}Y_{c33} + H_{c1}Y_{c32}Y_{c13} \end{pmatrix} v_{c2} + \begin{pmatrix} Y_{c11}Y_{c22}H_{c3} - Y_{c11}H_{c2}Y_{c23} - Y_{c21}Y_{c12}H_{c3} \\ +Y_{c21}H_{c2}Y_{c13} + H_{c1}Y_{c12}Y_{c23} - H_{c1}Y_{c22}Y_{c13} \end{pmatrix} v_{c3}}{\begin{pmatrix} Y_{c11}Y_{c22}Y_{c33} - Y_{c11}Y_{c32}Y_{c23} - Y_{c21}Y_{c12}Y_{c33} \\ +Y_{c21}Y_{c32}Y_{c13} + Y_{c31}Y_{c12}Y_{c23} - Y_{c31}Y_{c22}Y_{c13} \end{pmatrix}} \tag{2.18}$$

2.4.2 Prediction of the sound pressure from the operational velocity: derivation through the use of the coupled blocked force

2.4.2.1 Blocked force of coupled system

It is sometimes argued that structure-borne sound sources cannot be correctly treated by representing them by blocked force or free velocity because the source activity changes depending on the coupling conditions. Because of this coupling-variance it is desirable to obtain the blocked force in the coupled state. This subject has been extensively investigated by Moorhouse et al. [30]. They derived an expression for the coupled blocked force as:

$$\mathbf{F}_{bc} = \mathbf{Y}_c^{-1}\mathbf{v}_c. \tag{2.19}$$

The equation shows that the blocked force can be calculated from the coupled mobility and the operational velocities. By obtaining the coupled mobility with a direct measurement in the coupled state any influence from independent quantities is avoided. For a coupling-invariant source $\mathbf{F}_{bc}$ should of course be identical to $\mathbf{F}_b$. The terminology 'coupled' is used instead of '*in-situ*' because it better represents its origin.

A promising extension is presented in [30] by measuring at remote points away from the contact points. A large amount of points improves the accuracy and in practice it is often difficult to excite at the contact points. The remote method has two clear advantages in this respect.

2.4.2.2 Coupled transfer paths

In some cases it is desirable to predict the sound pressure out of measurements in the coupled state. By substitution of Eq. 2.4 in 2.8

$$p = \mathbf{H}(\mathbf{Y}_s + \mathbf{Y}_r)^{-1}\mathbf{Y}_s\mathbf{F}_b \tag{2.20}$$

and by introducing the coupled transfer path, $\mathbf{H}_c$, the sound pressure can be written as

$$p = \mathbf{H}_c\mathbf{F}_b. \tag{2.21}$$

By comparing Eq. 2.8 and 2.21 and by looking at the mechanical model in Fig. 2.2 the similarities between the conventional transfer path and the coupled transfer path can be seen. From those relationships it is also clear that the coupled transfer path can be obtained in a similar way. For the coupled case the *coupled system* has to be excited at one contact point at the time. Exactly there where the blocked force would be applied by the source. In this way the external excitation is consecutively providing a blocked force as if it were the internal activity of the source.

Now, by combining Eq. 2.19 and 2.21 the sound pressure can be predicted from the coupled transfer paths, the coupled mobility and the operational velocities

$$p = \mathbf{H}_c\mathbf{Y}_c^{-1}\mathbf{v}_c. \tag{2.22}$$

This matrix notation is equivalent to the result obtained in Eq. 2.18.

As a side note it is pointed out that Eq. 2.22 is closely related to the operational transfer path analysis (OTPA) known from NVH applications. OTPA estimates the factor $\mathbf{H}_\mathrm{c}{\mathbf{Y}_\mathrm{c}}^{-1}$ from the operational velocity and sound pressure records over time. A major drawback of OTPA is the fact that the method fails if strong correlation exists between the contact points. This is however usually the case for structure-borne sound sources. The measurement of coupled transfer functions and the implementation of Eq. 2.22 would in many cases lead to more reliable results compared to OTPA. Apart from an additional impulse hammer the setup would not have to be altered.

3 Importance of normal components: case study on a scale model

This chapter contains material from [31].

3.1 Introduction

This chapter presents a series of measurements conducted on a 1:4 scale model. The structure-borne sound source is realised by means of a rotational source with an out-of-balance mass. Scale model measurements allow for changes on the setup to be made with a reduced effort compared to measurements on a 1:1 scale and are therefore very well suited during the test phase of an investigation. However, it would require a tremendous amount of effort to build a scale model that represents a true mechanical system. First of all it would not be easy to accurately build a scale model of a timber joist floor or a washing machine and secondly, the laws that permit the translation from the scale model to the normal scale are not straightforward for structure-borne sound [32]. Nevertheless the measurements on the scale model revealed difficulties that were also important for the full scale measurements in chapter 4:

- The reproducibility of the small electric motor was low due to the quality of the bearings. Sleeve bearings lead to different behaviour depending on the play inside the bearing, the amount of oil, the rotational frequency and the different structures it is coupled to. The measurement procedure in this chapter is based on the simultaneous measurement of sound pressure and acceleration and the validity does not depend on the reproducibility of the source. However, changes in for example rotational frequency or harmonic content due to the state of a bearing have an influence on the results in third octave bands.

- Measurements of lightly damped structures that radiate in a receiving room with a long reverberation time are associated with very sharp resonance and anti-resonance peaks in the frfs. The predicted third octave band spectra depend on those sharp resonances and are very prone to errors. Common building materials involve considerable damping and it will in general be easier to make predictions with a higher accuracy in that case.

In this chapter the sound pressure in the receiving room is predicted out of the normal velocity at the contact points and a set of frequency response functions measured between the contact points and between the contact points and the microphone position. The velocity is measured in the normal direction to the plate during source operation. The frfs are measured on the coupled system at rest by exciting the plate in the normal direction. By doing this the remaining five degrees of freedom are simply omitted. The sound pressure is then predicted from the operational velocity and the measured frfs and it is compared to the directly measured pressure to investigate the influence of non-normal components on the radiated sound.

3.2 Measurement setup

The measurement setup is an attempt to roughly represent a 1:4 scaled living room excited at a timber floor ceiling with a 1:4 scaled washing machine. It is only a dimensional scaling and it was not intended to be a perfect scale model that behaves exactly like the true system.

A washing machine usually spins at 500-1600 rpm (8-27 Hz). For this reason, the rotational speed of the scaled washing machine is varied between 1800 and 8400 rpm (30-140 Hz). The pressure and the velocity in the scaled model are measured between 20 and 4000 Hz to cover the full scale frequency region of 5 Hz to 1 kHz. By using this frequency region the fundamental frequency of most rotational structure-borne sound sources and the acoustically interesting region are covered.

The source consists of an electric motor with a rotating disc connected to an

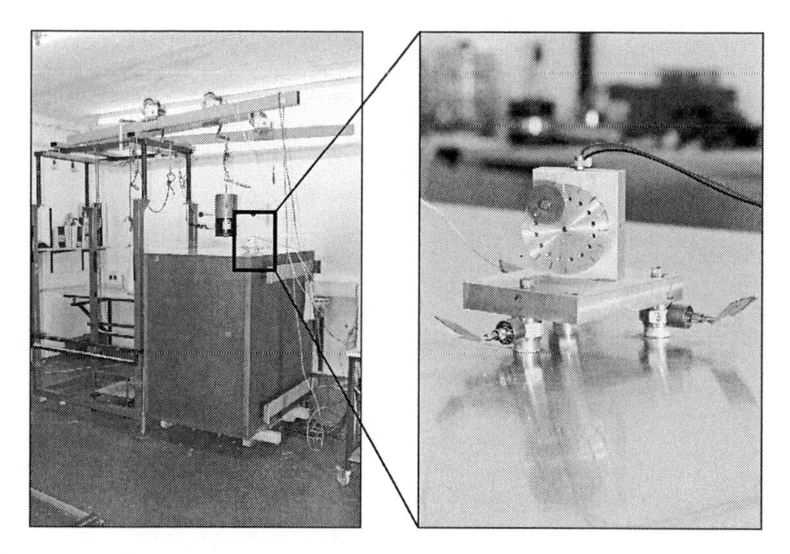

Figure 3.1: Scale model of the receiving room and the structure-borne sound source.

aluminium plate as shown in Fig. 3.1. The dimensions of the base plate of the scaled structure-borne sound source are 80 × 100 × 15 mm^3. Three force transducers are connected to the base plate and resemble the feet of the structure-borne sound source (these force transducers are not necessary for the measurement procedure and were only used for verification purposes). The feet are glued onto the receiver plate through aluminium cylinders with a diameter of 10 mm. The velocity is measured underneath the contact points with transducers attached to the plate. A series of out-of-balance masses can be mounted onto the rotating disc to vary the degree of source excitation.

The receiver plate is a Medium Density Fibreboard (MDF) plate of size $0.008 \times 1 \times 0.75\,m^3$. The plate is clamped between the edges of the receiving room and the weight of a metal frame. The receiving room is a box made out of 40 mm thick MDF plates. The dimensions are $1.25 \times 1 \times 0.75\,m^3$.

3.3 Scale model measurements

3.3.1 Verification with shakers

The theory in section 2.4 is first of all verified with three shakers mounted on the plate through stingers. This setup serves as a test case to verify the theory because the stingers provide an almost perfectly normal excitation. The predicted pressure based on the normal components should in this case be equal to the measured sound pressure due to the normal excitation.

The internal mass of the shakers is increased by an additional mass of 240 g to make them into non-ideal sources, c.f. Fig. 3.2 for a comparison of the source (shakers) and receiver (plate) mobilities. The frequency response functions were measured with the shakers in place according to Eq. 2.15. Without changing the setup, the shakers are subsequently fed with two different signals: sine tones at third octave band centre frequencies and correlated noise on all shakers. The pressure is calculated using Eq. 2.18 and compared to the *in-situ* measured pressure at a fixed position in the receiving room. Fig. 3.3 shows the ratio of *in-situ* measured pressure and calculated pressure. An almost perfect correlation is obtained for both excitation signals. The slight deviations of 0.5 dB are due to an imperfect normal excitation of the shakers.

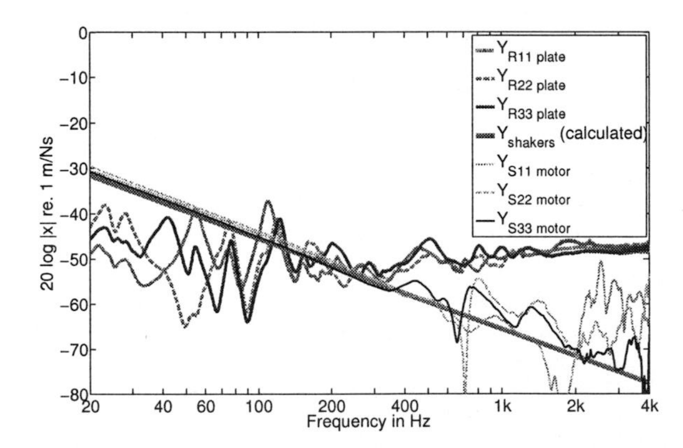

Figure 3.2: Comparison between plate and shaker mobilities.

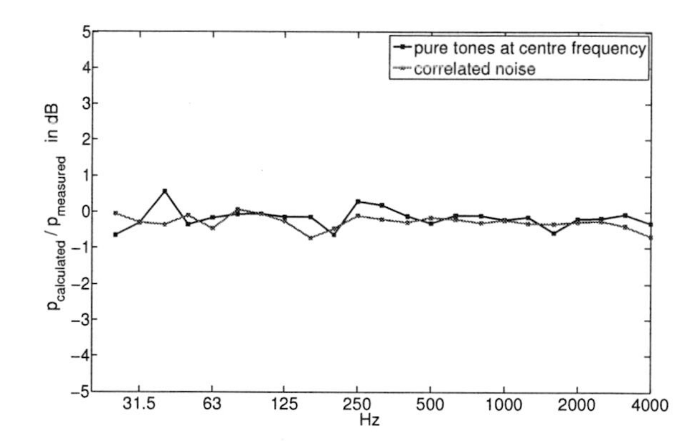

Figure 3.3: Comparison between the calculated pressure and the *in-situ* measured pressure for shaker excitation at three points on the plate.

3.3.2 Investigation with excentric motor

The measurement theory is now applied to the scaled structure-borne sound source depicted in Fig. 3.1. The comparison of the *in-situ* measured pressure and the calculated pressure will reveal to what extent this source can be characterised by its normal components. Any deviation from perfect agreement between measured and calculated pressure is attributed to the contribution of non-normal components.

Source characterisation is most complicated for cases where the source and receiver mobilities are of the same order of magnitude. Although the current chapter does not deal with an actual source characterisation it was still chosen to work with a system that represents this kind of coupling situation as seen in Fig. 3.2. However, it should be mentioned that the measurement procedure can be useful for all possible coupling situations.

For the following measurements the scaled source is excited with a B&K 4809 shaker to obtain the frequency response functions according to Eq. (2.15). Subsequently, the motor is run at different rotational speeds to measure the *in-situ* pressure and velocities. The pressure is then calculated according to Eq. (2.18).

3.3.2.1 Motor run at individual rotational speeds

The source is run at different rotational speeds to excite the system at different frequencies. The measured pressure spectra of all measurements are shown in Fig. 3.4. The spectrum of a single pressure calculation is compared to the *in-situ* measured pressure in Fig. 3.5. Besides the predominant fundamental frequency that corresponds to the rotational speed, a series of harmonics is transmitted to the sound field.

The full set of third octave band pressure ratios for different rotational speeds is shown in Fig. 3.6. Three different out-of-balance masses were measured at seven rotational speeds. Only levels 10 dB above the background noise were considered in the analysis. The average deviation (thick solid line) is ±1 dB at low frequencies. Above 2 kHz the difference is more than 5 dB.

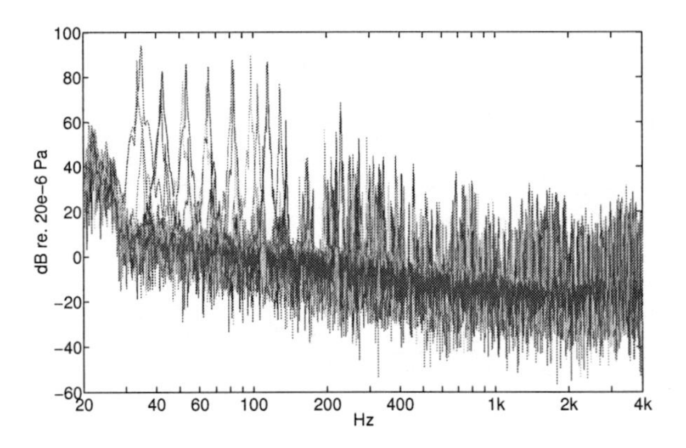

Figure 3.4: Measured sound pressure for different rotational speeds covers the frequency range between 30 and 135 Hz.

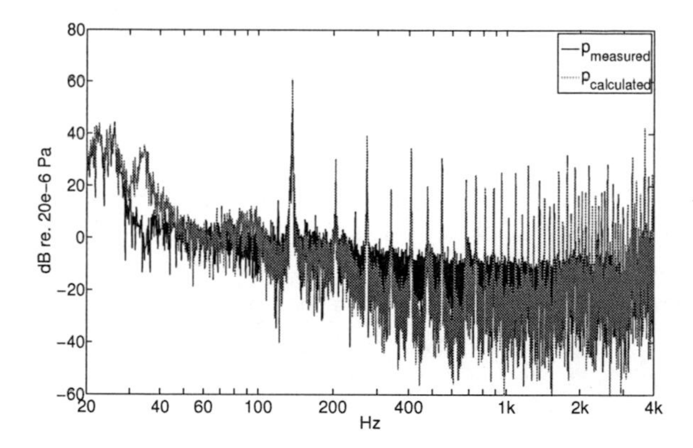

Figure 3.5: Comparison between the calculated pressure and the *in-situ* measured pressure for one rotational speed.

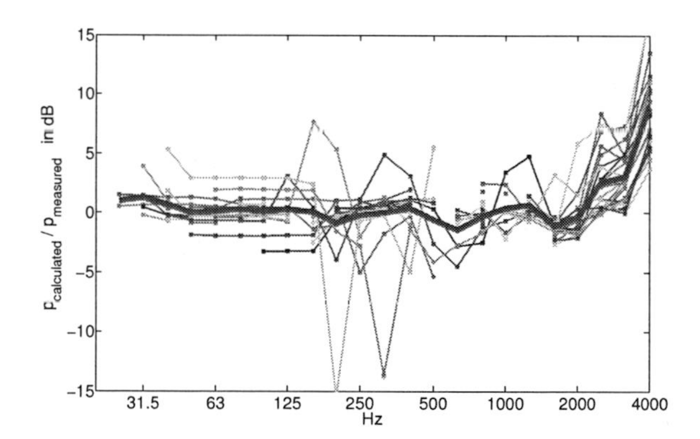

Figure 3.6: Comparison between the calculated pressure and the *in-situ* measured pressure for seven rotational speeds and three out-of-balance masses. The thick solid line is the average of all curves.

3.3.2.2 Motor run-up between 30 and 135 Hz

A broadband excitation can be achieved by measuring a run-up of the excentric motor. This is a convenient way to obtain an average result of the single rotational speed measurements in the previous section. By steadily increasing the fundamental frequency, the region between 30 and 135 Hz is well excited and the harmonics are smeared out due to the varying fundamental frequency. Fig. 3.7 shows the comparison of the measured and calculated pressure spectra and the deviation in third octave bands. The result is in agreement with the average of the single speed measurements shown in Fig. 3.6.

3.3.2.3 Measurement uncertainties

The measurement uncertainties involved in the measurement of the pressure ratio for one fixed position of the source on the plate is dealt with in this section. It directly applies to the measurements presented in the previous sections. For one such measurement the source is connected to the plate and is not disassembled at any time. Subsequently the frequency response function is measured followed by the *in-situ* velocities during source operation. The measurement uncertainty of the *in-situ* velocity measurements is assumed to be negligible. The uncertainty of the frequency response measurement according to Eq. (2.18) is investigated by repeating the measurement four times. The shaker is disconnected, repositioned and reconnected for each of the twelve measurements in total. The pressure ratio is calculated four times by using a single *in-situ* velocity measurement for a rotational speed of 7200 rpm. The result of the analysis is shown in Fig. 3.8. For most third octave bands the standard deviation is 0.3 dB.

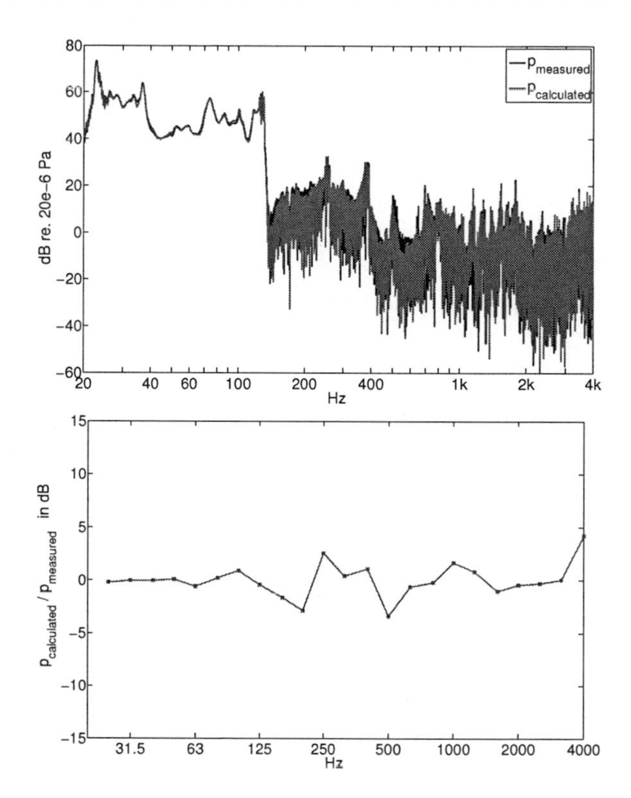

Figure 3.7: Comparison between the calculated pressure and the operational sound pressure for a swept excentric motor excitation between 30 and 135 Hz.

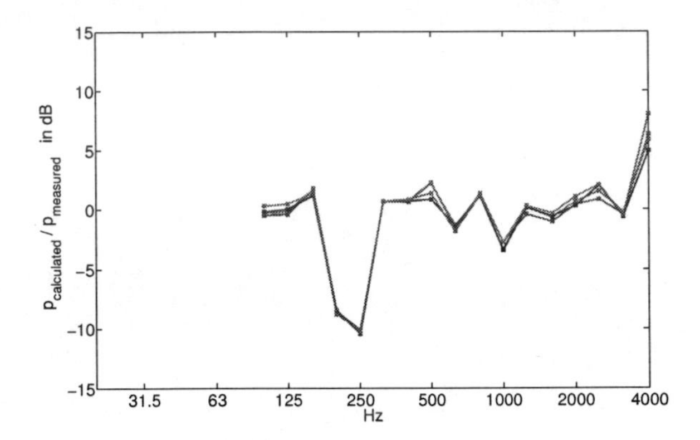

Figure 3.8: Measurement uncertainty of the frequency response measurement determined by repeating the measurement of the pressure ratio four times.

3.3.2.4 Measurement conclusions

From the last three measurement sections it can be concluded that this particular scaled source can be characterised by its normal components within ± 1 dB at the fundamental frequency. Above the fundamental frequency deviations of ± 3 dB are possible. Above 2 kHz the method shows differences of more than 5 dB. The standard deviation for those measurements is 0.3 dB. This means that a characterisation below 2 kHz of the scaled source by using independently measured normal components of source and receiver would result in a radiation prediction that differs by not more than ± 3 dB from the true radiation. This statement assumes that the characterisation itself is perfectly accurate. Considering that the source is rigidly connected to the plate through aluminium discs with a diameter of 10 mm, the accuracy is better than expected. Furthermore the error due to the simplification to the normal components has to be seen in the perspective of a complete source characterisation. The errors involved in an actual coupling prediction of a multipoint connected source are known to be much larger. This will be investigated in chapter 5.

3.4 Conclusion

It was shown that it is possible to precisely calculate the radiation of a scaled structure-borne sound source by using the *in-situ* operational velocity at all contact points and a series of frequency response functions. By measuring only a reduced set of components of the degrees of freedom at the excitation, the predicted pressure only represents this simplified approach. By then comparing the predicted pressure with the *in-situ* measured pressure, information is revealed on how well a source-receiver characterisation through independently measured source and receiver mobilities based on those simplifications will work.

Measurements on a scale model of a structure-borne sound source showed that the predicted and measured pressure correlate very well. On average deviations of 3 dB were found. In the next chapter the same investigation is conducted on a full scale wooden joist floor with a washing machine.

4 Importance of normal components: a case study of washing machine

This chapter contains material from [27].

4.1 Introduction

The same investigation presented in the previous chapter is conducted in this chapter on a full scale wooden joist floor and a normal washing machine. Again the predicted sound pressure level is calculated from the operational normal velocity and the measured frfs between the contact points and the microphone position. The calculated sound pressure is compared to the directly measured sound pressure to reveal the influence of the non-normal components. Two additional aspects are looked at: linearity and the degree of moment excitation.

The linearity of the complete system is investigated by increasing the excitation through the use of different out-of-balance masses in steps of 20 g. The comparison between calculated and directly measured pressure should not depend on those variations.

The degree of moment excitation is increased by increasing the contact area of the feet. Large feet are able to transfer moments more easily due to a large moment arm. A resilient layer between the feet and the floor was used to decrease the moment excitation.

(a) receiving room

(b) wooden joist floor

(c) ideal source

(d) washing machine

Figure 4.1: Measurement setup with two vibrational sources connected through three feet to the wooden floor.

4.2 Measurement setup

The measurements were conducted on a wooden joist floor in a transmission suite according to ISO 140-11 as shown in Figure 4.1. The 20×10 cm wooden joists are covered with 25 mm thick chipboard plates with tongue and groove joints. The size of the floor is 6.5×4 m. The receiving room has a volume of 80 m^3 and a reverberation time of 0.9 s.

The sources depicted in Figure 4.1 are used in the experiments. The ideal source consist of an electrical motor of a washing machine that is fixed onto a chipboard base plate (50×40×5 cm, W×D×H). It is ideal in the sense that it does not contain non-linear springs and dampers in combination with large amplitudes.

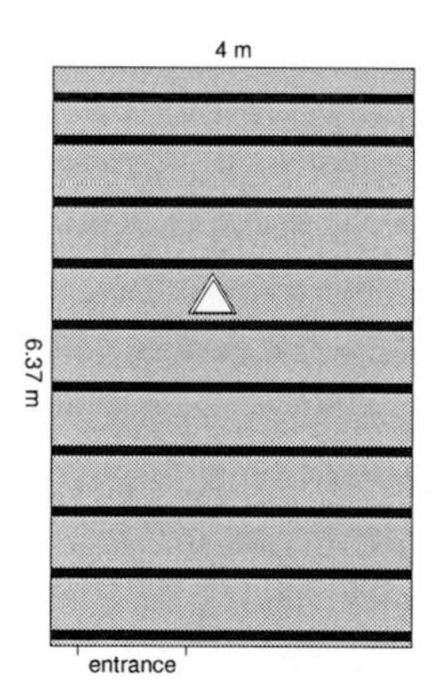

Figure 4.2: The position of the sources on the floor. The sources are positioned between two joists.

The mobility of the separate electrical motor will be dominated by its mass-like behaviour [33] and can be treated as a linear component. The vibration amplitudes of the wooden base plate are small enough to assume linearity. The three feet are rigidly connected to the base plate with bolts and nuts.

The second source is a standard washing machine (60×50×85 cm, W×D×H) with chipboard plates screwed onto its sides to provide additional damping of the metal sheets. The front feet are removed and replaced by a single foot in the middle to obtain a source with three feet and avoid problems of aligning the fourth foot. Figure 4.3 depicts the different types of feet that were chosen for the experiments.

The source excitation is varied by increasing the out-of-balance mass in 20 g steps up to 200 g. The masses are attached to the rotating disc of the ideal source and to the drum of the washing machine. This resembles the excitation caused by laundry inside a washing machine. Laundry will generally have a lower out-of-balance effect because it is evenly spread along the wall of the drum due to a slowly increasing spinning speed as it is programmed in a modern washing machine. The maximum out-of-balance mass of 200 g as it is used in the experiments exceeds the practically relevant excitation levels.

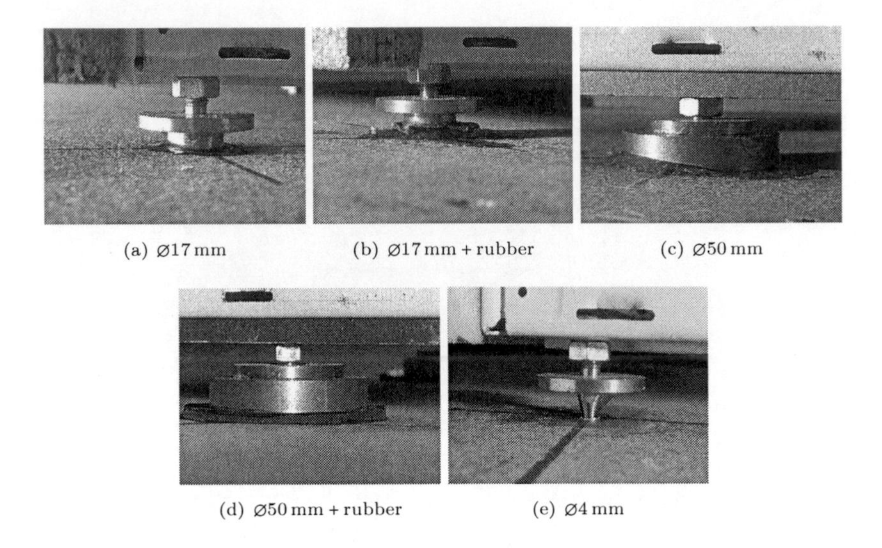

(a) ⌀17 mm (b) ⌀17 mm + rubber (c) ⌀50 mm

(d) ⌀50 mm + rubber (e) ⌀4 mm

Figure 4.3: Feet with different diameters used on the structure-borne sound sources shown in Figure 4.1. Feet 4.3(a)-4.3(d) are glued with epoxy glue, foot 4.3(e) is screwed with a bolt of 3 mm in diameter.

4.3 Measurement procedure

The ideal source and the washing machine are driven by a harmonic-like excitation by controlling the rotational speed. The sources are run up in 24 seconds from 16 to 20 Hz and from 4 to 20 Hz, respectively. A custom made unit to control the rotational speed of the electrical motor is controlled by the measurement software and produces perfectly reproducible run-ups. By running up the motor a broadband response is obtained. Any errors that occur in the measurement procedure as it is described in the following paragraphs are averaged out by the broadband response. Measuring at a single excitation frequency and its harmonics is very prone to errors because the band levels are determined by only a few frequencies. In chapter 3 it was shown that the average of measurements at single frequencies is comparable to the result of a run-up. In practice an average broadband result is usually more meaningful than a response at a single frequency.

During the run-ups the normal *in-situ* velocity at the three feet is measured underneath the wooden floor. The sound pressure in the receiving room is measured at five microphone locations simultaneously and denoted by $p_{x\,\text{measured}}$, the *in-situ* sound pressure at position x.

After that the source is switched off and left in place to measure the frequency response functions between the feet and the microphone positions. This is done by using an impulse hammer to excite the contact points through the wooden floor. From the normal *in-situ* velocities and the frequency response functions the sound pressure at position x is calculated according to Equation 2.18 and denoted by $p_{x\,\text{calc. from normal}\,v}$.

The goal of the current investigation is to determine whether the radiated sound available from $p_{x\,\text{measured}}$ deviates from a prediction that only accounts for the normal excitation components as it is calculated in $p_{x\,\text{calc. from normal}\,v}$. Most important is to find out how strong the difference will be perceived by the human hearing. This fact justifies the analysis in third-octave bands. Loudness was not used because of the relatively small level differences over the frequency range and hence the limited influence of masking. The sound pressure levels in dB(A) in the receiving room during excitation by the washing machine can be seen in Figure 4.7(a). The frequency range of the analyses is limited from 20 to 1000 Hz because this is generally the most important region for structure-borne sound in buildings. The sound pressure levels in the following experiments are not entirely comparable to real-world situations because the absorption material and the common gypsum board ceiling have been left away for practical reasons. The acoustical annoyance from rotating structure-borne sound sources like washing machines will generally be dominant below 500 Hz.

From the measured and calculated sound pressure third-octave band levels are obtained. Levels with an SNR below 10 dB are not evaluated. The ratio of the band levels is denoted by $p_{x\,\text{calc. from normal}\,v}/p_{x\,\text{measured}}$. To improve the result an average over five microphone positions is taken. The average sound pressure ratio over five microphones is denoted by $\langle p_{\text{calc. from normal}\,v}/p_{\text{measured}}\rangle$.

4.4 Verification with a point-connected ideal source

To verify whether the theory works under ideal conditions the ideal source shown in Figure 4.1 was used as a structure-borne sound source. To provide a test case with negligible moment excitation the connection area between the source and the wooden floor was reduced to a minimum by using feet that resemble point connections as shown in Figure 4.3(e). The diameter of the connecting surface of the point connections is 4 mm. The steel cones accommodate a 3 mm bolt to screw the feet to the floor. This makes sure that the source stays in place at high excitation levels. Clearance between the feet and the floor during operation would result in a non-linear behaviour of the system. This is something that cannot be dealt with by the mobility approach and has to be avoided in the experiments. Due to the characteristics of the source and the way the source is connected to the floor this test case will most likely behave as a linear system with negligible moment excitation. In this case $p_{x\,\mathrm{measured}}$ and $p_{x\,\mathrm{calc.\,from\,normal}\,v}$ should ideally be identical.

In the first verification measurement, the ideal source was externally excited with a shaker from the top to provide a broadband excitation. As a consequence of using this shaker orientation the predominant excitation will occur in the normal direction to the floor and will additionally reduce moment excitation at the feet. The broadband calculated and measured sound pressure is shown in Figure 4.4(a). The agreement is very good. The third-octave band levels in Figure 4.4(b) are very close to 0 dB which means that the radiation from the source under test can be almost perfectly predicted by only considering the normal components. This is not surprising for a point-connected source under normal excitation.

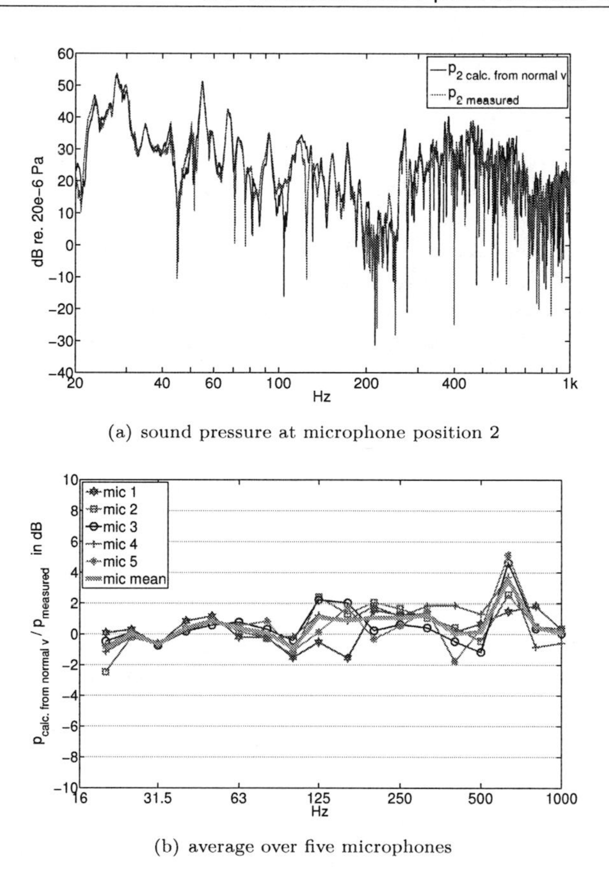

Figure 4.4: Ideal source excited with an external shaker in the normal direction to the floor.

4.5 Verification with a point-connected washing machine

The same verification process was repeated for the washing machine. The washing machine was point-connected to the floor at the same position. To obtain the measurement uncertainties of the following measurements the reproducibility and the repeatability was investigated. The washing machine was connected to the floor and the frequency response functions and the *in-situ* velocities were measured three times. This process was repeated three times after disconnecting the washing machine and reconnecting it at the same position to the floor. The results of the measurement uncertainty analysis for the normal external shaker excitation is shown in Figure 4.5(a). The standard deviation for most of the third-octave bands is below 1 dB with a maximum value of 1.7 dB. The mean value of the sound pressure ratios is close to 0 dB with deviations up to ± 3 dB. This shows that the theory works quite well but not perfectly for the point-connected washing machine under normal excitation.

More interesting is the behaviour of the washing machine during normal operation. A run-up between 4 and 20 Hz was investigated with an out-of-balance mass of 180 g in Figure 4.5(b). The measurement was also repeated nine times as explained in the previous paragraph. The mean curve of the sound pressure ratios is close to 0 dB with deviations up to ± 2 dB. The measurement uncertainties expressed as a standard deviation are below 1 dB. This proves that the point-connected washing machine can be characterised by its normal components if the moderate error shown in Figure 4.5(b) is acceptable.

The measurement uncertainties shown in Figure 4.5(b) are representative for the results of the washing machine in section 4.6 and 4.7.

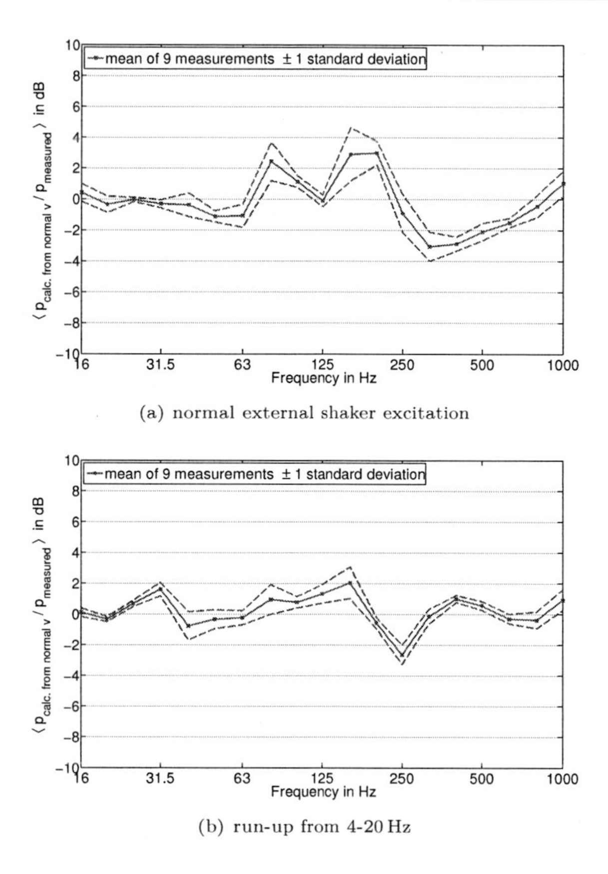

(a) normal external shaker excitation

(b) run-up from 4-20 Hz

Figure 4.5: Sound pressure ratio of nine measurements of a point-connected washing machine to obtain an estimate of the measurement uncertainties.

4.6 Linearity

To investigate the linearity of the sources coupled to the floor the excitation level was increased by varying the out-of-balance mass in steps from 0 g to 200 g. In a linear system there should be no change in sound pressure ratio ($p_{x\text{ calc. from normal }v}/p_{x\text{ measured}}$) depending on the out-of-balance mass. This is tested in the following section for the ideal source and the washing machine. The maximum weight of 200 g was determined by observing the movement of the washing machine on three rubber feet. The mass was increased and the whole machine started moving across the floor just above 200 g. Any movement of the feet across the floor necessarily involves clearance between the feet and the floor and introduces non-linear behaviour. The same maximum mass was also used for the ideal source. However, because of its lower weight, the source starts moving around at much lower out-of-balance masses. For this reason all linearity experiments were conducted by fixing the feet onto the floor. To avoid the influence of moment excitation the point connections out of Figure 4.3(e) were chosen.

The sound pressure range covered during a run-up of the ideal source by the different out-of-balance masses is shown in Figure 4.6(a). The excitation of the motor run-up between 16 and 20 Hz can clearly be seen in the sound pressure level. Those frequencies are acoustically less relevant but they correspond to the normal spinning frequencies of a washing machine. The audible part consists of the harmonics of the run-up. The corresponding velocity during this run-up is shown in Figure 4.6(b). The limiting factor in the experiment was the very high velocity level in the 20 Hz band that led to considerable floor displacements. Comparison with realistic velocity levels of a washing machine in Figure 4.7(b) shows a difference of about 15 dB.

The results for the linearity analysis for the point-connected ideal source are shown in Figure 4.6(c). For the out-of-balance masses between 0 and 80 g the differences between the curves are within 1 dB. This shows that the source behaves almost perfectly linearly within this range. Above 80 g the curves around 50 Hz are 6 dB higher. This is an indication that those high velocity levels lead to non-linearities for the ideal source.

More important is the linearity analysis of the standard washing machine. A washing machine has a built-in vibration isolator and therefore produces lower sound pressure and velocity levels as shown in Figure 4.7. The differences between the levels for the different out-of-balance masses are also reduced to a range of about 6 dB. Figure 4.7(c) shows the sound pressure ratios for the point-connected washing machine. The difference between the ratios is within 1 dB with a stronger deviation at 250 Hz. Considering the measurement uncertainties there is no significant difference between the results for the different out-of-balance masses.

To further investigate the linearity boundaries of the washing machine on a wooden floor the maximum mass was increased up to 600 g. The analysis is shown in Figure 4.8. Interestingly the sound pressure ratios for the very heavy out-of-balance mass of 600 g show a similar linear behaviour. The difference between the curves is also within 1 dB. To conclude this section it can be said that the washing machine can be treated as a linear system for the excitation levels between 0 and 600 g.

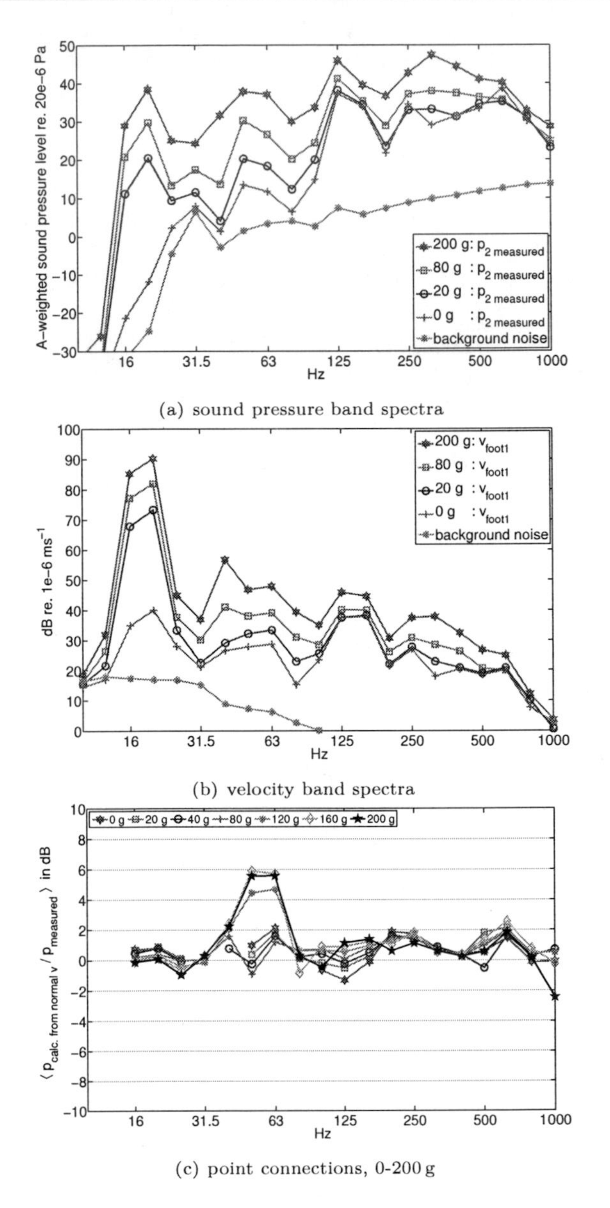

(a) sound pressure band spectra

(b) velocity band spectra

(c) point connections, 0-200 g

Figure 4.6: Linearity analysis of the ideal source on point connections according to Figure 4.3(e). The source is run up from 16 to 20 Hz with different out-of-balance masses.

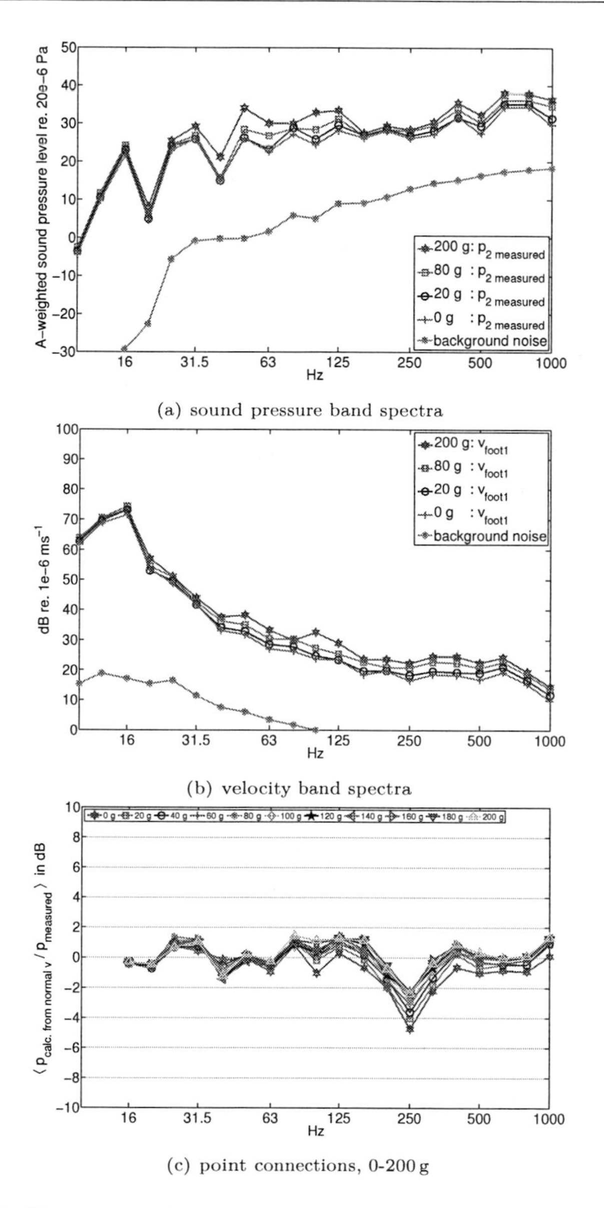

(a) sound pressure band spectra

(b) velocity band spectra

(c) point connections, 0-200 g

Figure 4.7: Linearity analysis of the washing machine on point connections according to Figure 4.3(e). The source is run up from 4 to 20 Hz with out-of-balance masses up to 200 g.

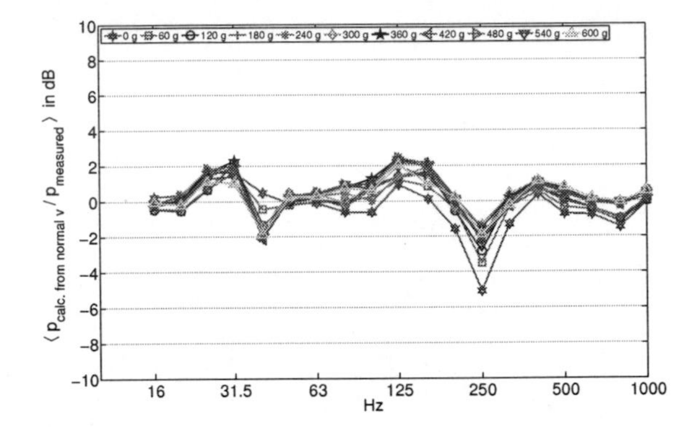

Figure 4.8: Linearity analysis of the washing machine on point connections according to Figure 4.3(e). The source is run up from 4 to 20 Hz with out-of-balance masses up to 600 g.

4.7 Moment excitation: variation of the contact area

The mobility theory treats contact areas between source and receiver as discrete points and does not account for the pressure distribution at the contact area. For this to be valid the contact area has to be small. A rule of thumb is given in [21]: the dimensions of the contact area should be smaller than a tenth of the bending wavelength. According to this prerequisite, the foot size does not matter as long as it is smaller than a tenth of the wavelength. However, in practical setups there is one important aspect that has to be considered: small feet with a small radius have a small moment arm and involve very high forces at the edge of the feet. Those forces are responsible for the moment excitation of the receiver structure. High forces lead to local plastic deformation at the contact area and will not transfer the moment. On the wooden floor for example, the surface of the chipboard plate would be irreversibly indented. In other words, a certain foot area is needed if the moment is to be transferred to the receiver structure in a practical setup. An extreme example of this is the ideal point connection in which the moment arm is zero which results in a vanishing moment transmission.

Figure 4.9 depicts this practical aspect for the z-component. The moment M_z equals the moment of inertia, J, times the rotational velocity, w_z, and will not change with foot size. The moment is transferred to the plate through a particular pressure distribution depending on the surface of the foot. For a flat foot in contact with a flat plate surface the pressure distribution will be at its maximum at the edge of the foot. This can be looked at in a simplified way by only assuming contact at the edge a distance r away from the centre through the force F_{M_z}. In that case the transferred moment M_z equals $r \cdot F_{M_z}$. For smaller feet the distance r decreases and the force necessarily has to increase to transfer the same moment.

In the previous paragraph the moment excitation at the separate feet was investigated. Those are the moments that are used in a multi-point mobility source characterisation. The moment of the washing machine as a whole is accounted for by the phase information between the different feet.

The distance between the feet of the washing machine can be related to the bending wavelength on the plate. For frequencies at which half of the plate bending wavelength equals the distance between the feet, a strong moment will occur because of the maximum height difference between the feet. For a distance

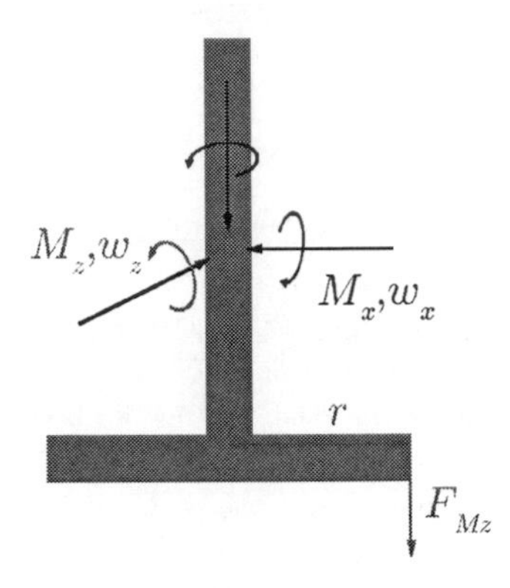

Figure 4.9: Moment excitation of the z-component for a foot with radius r.

of 53 cm for the washing machine this corresponds to a frequency of 350 Hz. For the ideal source the distance between the feet of 43 cm and 46 cm is reached at 530 Hz and 460 Hz.

The bending wavelength of a chipboard plate of thickness 25 mm with a density of 700 kg/m^3 and a Young's modulus of $3 \cdot 10^9\,N/m^2$ is greater than 0.3 m below 1 kHz. If the influence of the wooden joists on the bending wavelength is assumed to be negligible, feet with a diameter smaller than 30 mm can be assumed to be discrete point connections for the mobility theory up to 1 kHz.

The feet in the following experiment are 4 mm, 17 mm and 50 mm in diameter. The smallest diameter is an almost ideal point case and should transfer no moments. The 17 mm foot represents a normal sized foot of a washing machine. The 50 mm version was chosen as an example of perfect moment excitation. To provide good contact with the floor the point connections are screwed on to the floor. The other feet are glued with a dual compound epoxy glue to make sure that the contact layer is as stiff as possible. A thin resilient layer would easily absorb the high edge forces through its damping and result in a reduced moment excitation. Those three diameters are used to investigate the influence of moment excitation in the case of firmly connected feet.

The same 17 mm and 50 mm feet are then uncoupled by use of a soft rubber layer of thickness 3.5 mm. The rubber layer is especially meant to reduce the moment excitation but will also have an influence on the normal force. The rubber layer is glued onto the feet and onto the floor to avoid any movement of the washing machine across the floor. Those cases allow the investigation of moment excitation for a more realistic foot with a rubber interlayer.

A final example consist of 17 mm feet with rubber layer simply positioned onto the floor without glue. This case is the most realistic washing machine foot and was used to show how much moment excitation really matters in practice.

The described series of feet allow investigations ranging from no moment excitation (point connection) to perfect moment excitation (50 mm feet firmly connected). The ratio $p_{x\,\text{calc. from normal}\,v}/p_{x\,\text{measured}}$ is investigated in the following analyses for different excitation levels with out-of-balance masses varying between 0 and 200 g. If the moment excitation contributes to the sound radiation the sound pressure ratio will not be 0 dB.

4.7.1 Ideal source

The sound pressure ratios for different feet are shown in Figure 4.10. By comparing the results for point connections in Figure 4.10(a) with the firmly connected 50 mm discs in Figure 4.10(b) no significant differences can be found apart from the peaks around 63 Hz. Those peaks were previously discussed as being non-linearities due to the high velocity levels for the heaviest out-of-balance masses. Looking at the general average of 0 ± 2 dB in all graphs it can be concluded that the moment excitation is not greater for large feet. The sound pressure ratios for the 50 mm discs with rubber layer are almost identical to the result without the rubber layer. As a general conclusion about the ideal source it can be said that the source can be represented by its normal components with a reasonable accuracy of 0 ± 2 dB.

4.7.2 Washing machine

The sound pressure ratios for the point-connected washing machine were shown previously in Figure 4.7(c) and serve as a verification of the ideal case with negligible moment excitation. The results are on average 0 ± 1 dB with an outlying value at 250 Hz. The outlying value could be related to the distance between the feet of the washing machine. It was mentioned that half the plate bending wavelength is reached at 350 Hz.

By comparing this to the results for the 17 mm feet in Figure 4.11(b), a general tendency of negative ratios above 125 Hz is seen. This is an indication of stronger moment excitation at higher frequencies. Below 125 Hz the average ratio is 0 dB with a few distinct peaks. Those peaks are related to the moment of the complete washing machine and were identified by looking at the phase relation between the feet.

If a layer of rubber is added to the 17 mm feet in Figure 4.11(c) the deviations below 125 Hz disappear and the average is almost 0 dB. Above 125 Hz an average ratio of -3 ± 2 dB is seen. In this frequency range the expected improvement of a rubber layer on the moment excitation could not be found.

The results for the 50 mm feet show a similar behaviour. The ratios for the firmly connected discs below 125 Hz in Figure 4.12(a) are close to 0 dB with a very strong deviation at 31 Hz. This was again due to two feet being perfectly out-of-phase. Above 125 Hz an average ratio of -3 ± 2 dB is observed. The measurements with the same feet with an additional rubber layer in Figure 4.12(b) are very similar to the 17 mm feet with rubber. The improvement of the rubber layer is only significant below 125 Hz.

The final case is the 17 mm foot with an unglued rubber layer. This is the most relevant case for practical source characterisations. The curves in Figure 4.11(a) follow a similar trend to all other results. Below 125 Hz the ratios are almost 0 dB. Above 125 Hz an average ratio of -2 ± 2 dB is seen. This means that the washing machine can be characterised with good accuracy even if the degrees of freedom are reduced to the normal components.

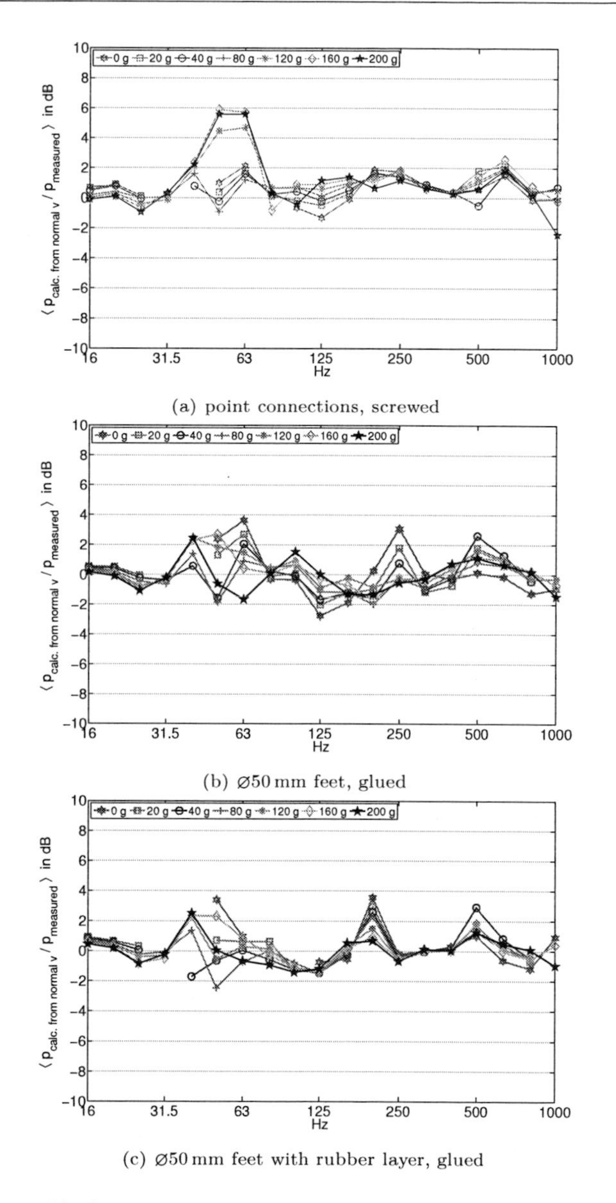

(a) point connections, screwed

(b) ⌀50 mm feet, glued

(c) ⌀50 mm feet with rubber layer, glued

Figure 4.10: Ideal source run with different feet and different out-of-balance masses. The motor is run up from 16 to 20 Hz. The out-of-balance mass is increased from 0 g to 200 g. The feet are according to Figure 4.3.

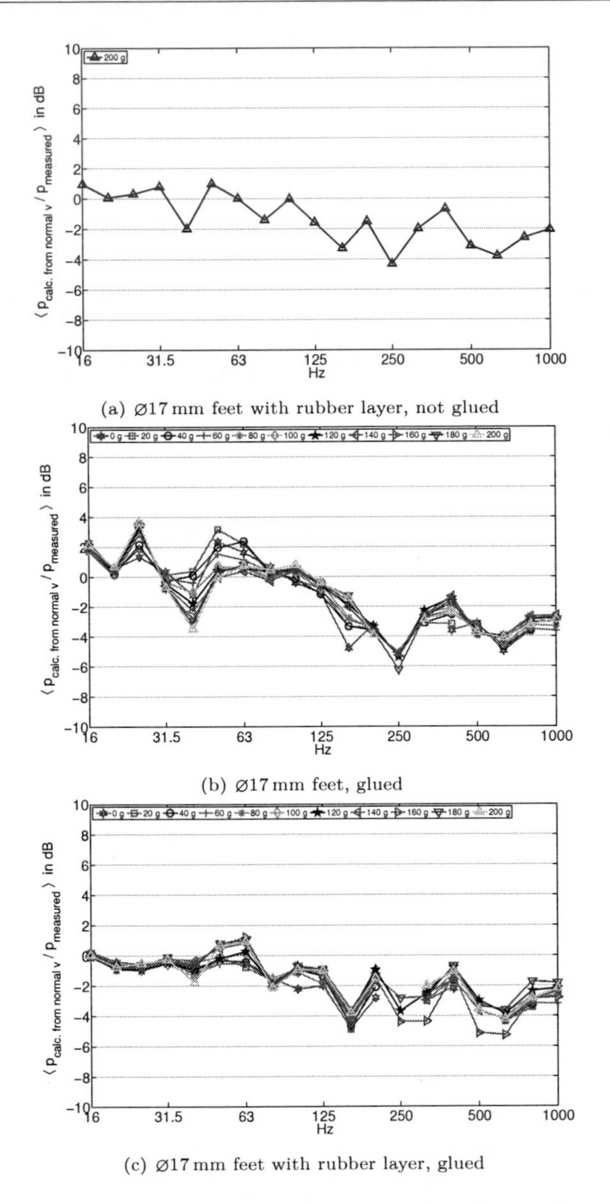

(a) Ø17 mm feet with rubber layer, not glued

(b) Ø17 mm feet, glued

(c) Ø17 mm feet with rubber layer, glued

Figure 4.11: Washing machine run with different feet and different out-of-balance masses. The motor is run up from 4 to 20 Hz. The out-of-balance mass is increased from 0 g to 200 g. The feet are according to Figure 4.3.

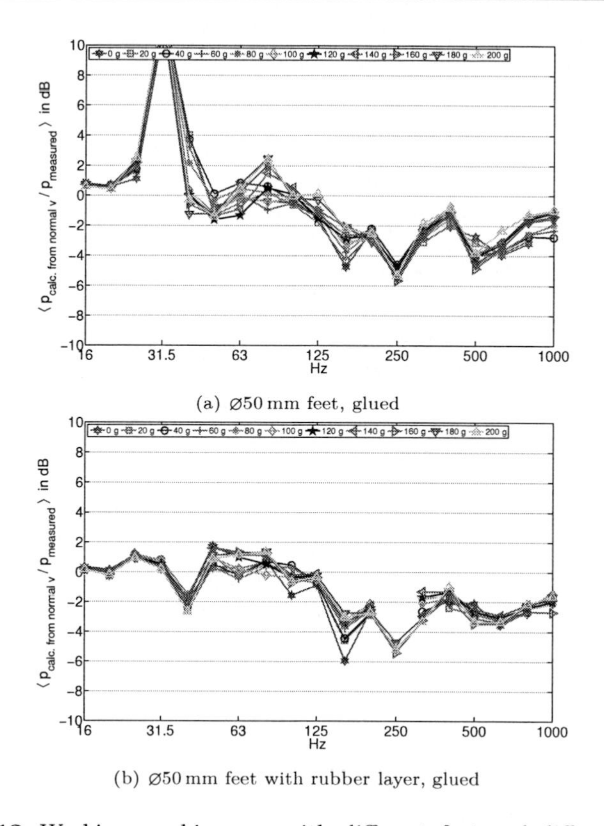

(a) Ø50 mm feet, glued

(b) Ø50 mm feet with rubber layer, glued

Figure 4.12: Washing machine run with different feet and different out-of-balance masses. The motor is run up from 4 to 20 Hz. The out-of-balance mass is increased from 0 g to 200 g. The feet are according to Figure 4.3.

4.8 Conclusion

A case study of a washing machine on a wooden floor was conducted with respect to its linear behaviour and its radiation due to non-normal components. The measurements revealed that the washing machine can be treated as a linear source on the wooden floor. The ratio of the calculated and measured sound pressure showed that the contribution due to non-normal components is only important for firmly connected feet. The error is on average 3 dB for frequencies above 125 Hz. Below 125 Hz deviations up to 10 dB were measured. For feet with a rubber layer the deviations are on average 2 dB above 125 Hz and 0 dB below 125 Hz. Those results represent a run-up of the source between 4 and 20 Hz.

The case study proves that a washing machine on a wooden floor can be represented by its normal components with a very limited loss of accuracy. This is especially true for a standard installation on rubber feet. Whether those findings apply to all vibrational sources in buildings is not answered. For sources with similar dimensions, excitation levels and contact situations it is likely to be the case. Further case studies with different sources and receiver structures will have to be performed to answer this question definitely.

5 Predicting the interaction: a case study of a washing machine

This chapter contains material from [26].

5.1 Introduction

This chapter is concerned with source characterisation in the true sense. A source is measured independently from the receiver structure and the quantities in the coupled state are predicted by means of calculation. The comparison between the prediction and the directly measured quantities in the coupled state is conducted in the sound field by comparing the sound pressure in both cases.

The prediction only involves the normal components. To reduce the likelihood of moment excitation the contact area is made as small as possible. It is assumed that the in-plane components (translational x and z-components) can be neglected.

5.2 Measurement setup

The measurements is very similar to the measurement setup in section 4.2. A few modification were made:

- The ideal source was modified to make its mobility more similar to the mobility of the washing machine. The source with size $60 \times 50 \times 35\,\mathrm{cm}$ (W×D×H) is depicted in Fig. 5.1.

- The diameter of the contact points is 9 mm for the washing machine and 4 mm for the ideal source as shown in Fig. 5.2.
- The out-of-balance mass of the washing machine and the ideal source is 180 g and 120 g respectively. It is kept constant throughout the measurements in this chapter. The property of linearity was not investigated in this chapter.
- The sources were positioned at three different locations on the floor according to Fig. 5.3.

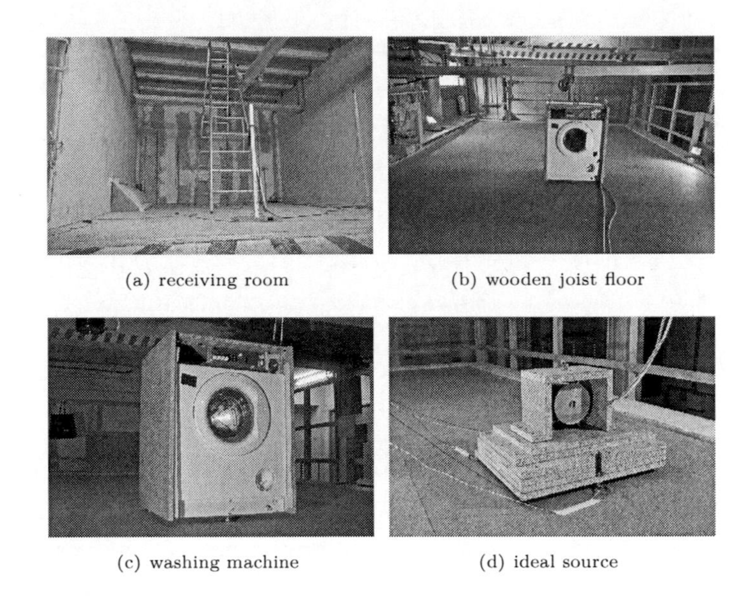

(a) receiving room (b) wooden joist floor

(c) washing machine (d) ideal source

Figure 5.1: The vibrational sources are connected through three feet to the wooden floor.

All measurements were performed in Matlab with the ITA-Toolbox [34]. Brüel and Kjaer force (Type 8200) and acceleration (Type 4397) transducers were used with Brüel and Kjaer Nexus pre-amplifiers. The DA conversion was done with an RME Multiface. KE-4 microphones by Sennheiser were used at fixed points in the receiving room and pre-amplified with an RME Octamic. The frequency response functions were measured with a sampling rate of 44.1 kHz and a sweep

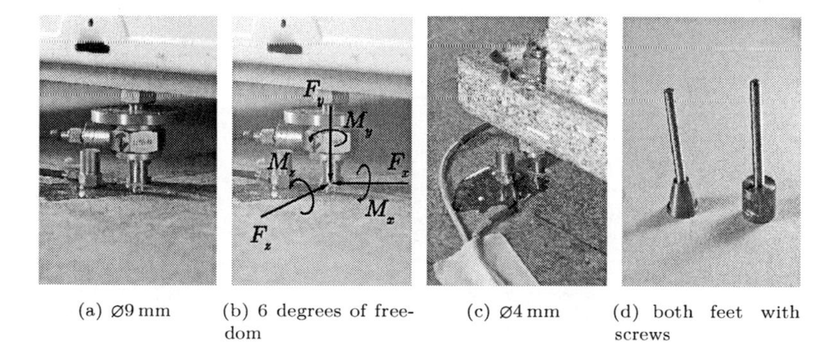

(a) Ø9 mm (b) 6 degrees of freedom (c) Ø4 mm (d) both feet with screws

Figure 5.2: Mounting conditions with feet with different diameters: 9 mm on the washing machine and 4 mm on the ideal source. The contact area is as small as possible to reduce the influence of moment excitation. In (d) the feet are turned upside down to clearly show the contact area. The screws are used to connect the feet to the chipboard plate.

length of 2^{21} samples. Averaging and windowing was used in most cases.

5.3 Measurement of independent quantities

The coupled state of the source is predicted with independently measured quantities: the free velocity, the blocked force, the source mobility and the receiver mobility. The setups for the different measurements are shown in Fig. 5.4. The measurements were conducted according to ISO 9611 and 7626 [35, 36]. The different measurement procedures on the basis of those standards are described in the following sections.

The measurement range was limited from 22 to 890 Hz as it is the most important frequency range for the sound radiation from structure-borne sound sources in buildings. The sound radiation above 1 kHz will generally be very low in common buildings with common building materials. If there is a need to predict and measure above 1 kHz the measurements become even more challenging. The

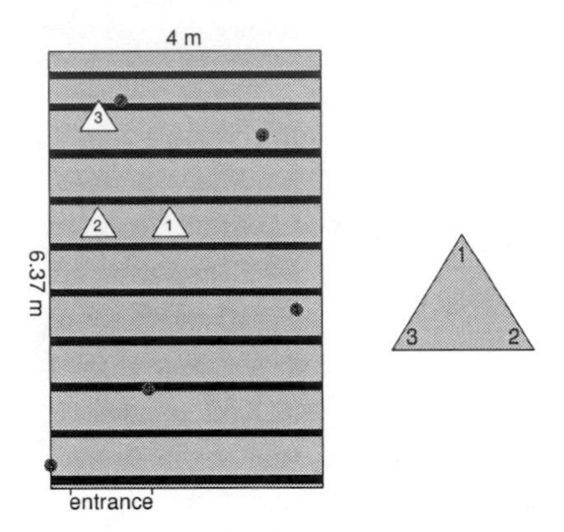

Figure 5.3: Three different positions of the sources on the floor. Contact point 1 is at the front of the source. The dots designate the position of the microphones in the receiving room. Top view, drawn to scale.

measurement uncertainty increases dramatically and the reproducibility depends on differences in the range of a few millimeters.

The velocity was measured with one accelerometer on-axis except for the coupled case where the accelerometer was mounted 1.5 cm off-axis. It was verified to be identical to the on-axis velocity for the frequency range of interest.

The reproducibility of the measurements is important because the analyses in the section 5.4 will use certain quantities to predict the coupled state. The accuracy of the prediction depends of course on the reproducibility of the separately measured quantities.

It is known that the source activity of structure-borne sound sources is often instable in terms of reproducibility. However, very little information concerning this aspect has been collected in the past. The poor reproducibility of rotational sources is mainly attributed to the fact that the spectrum of the source activity consists of a fundamental frequency (according to the rotational speed) and additionally a series of harmonics. The fundamental frequency is a linear property but the harmonics are strictly speaking non-linearities. Those non-linearities tend to exhibit chaotic behaviour and result in a low reproducibility. A perfectly

(a) free velocity (b) source mobility

(c) blocked force (d) receiver mobility

Figure 5.4: Different measurement setups to obtain the independent quantities.

balanced rotation in a linear system should result in a motion at a single frequency without any harmonics. However all practical rotational sources will produces harmonics and the audible frequency range might even be above the fundamental frequency.

The reproducibility of the free velocity spectrum at a single rotational frequency was investigated for the washing machine and the ideal source. The third octave band levels vary by up to ±10 dB. A verification of a prediction is in that case only possible through many averages of reproduced setups. To avoid those averages a run-up of the source was tested. The rotational frequency of the source is slowly increased from 1.2 Hz to 19 Hz in 47 seconds. The final 23 seconds are recorded. The reproducibility of the run-up of the different quantities proved to be much better and will be discussed in the corresponding sections.

Another crucial aspect is the invariance of the source behaviour depending on different coupling conditions. It is investigated whether the harmonic content depends on the coupling situation. If a source is going to be characterised with a linear system approach the frequencies of the different harmonics should not change for different coupling situations otherwise the characterisation is not valid. This was observed in a case study on a scale model in [37]. The harmonic content of the washing machine was investigated in three different situations: coupled to the floor, blocked and freely suspended as shown in Fig. 5.5. It can be seen that the frequencies of the harmonics are the same for the different coupling situations. This prerequisite, the coupling-invariance of the source activity, is also fulfilled for the ideal source.

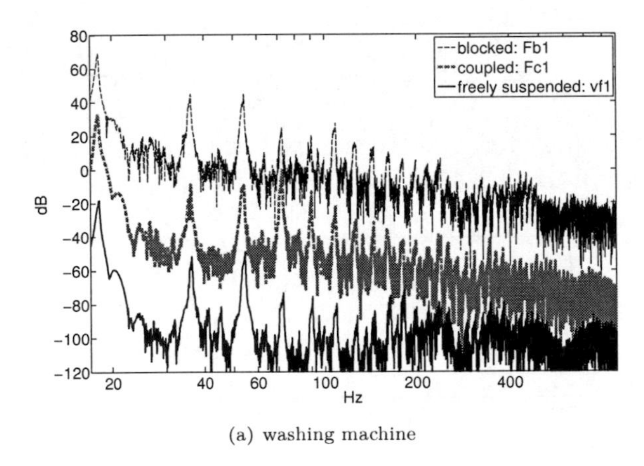

(a) washing machine

Figure 5.5: Comparison of the harmonic content for different states of the washing machine: coupled to the floor, blocked and free. The curves are shifted to simplify the comparison.

5.3.1 Free velocity

The measurement of the free velocity is the most straight-forward of all measurements. The free state demands that the source is measured during source operation without any external force being applied to the source. This can be achieved by suspending the source on springs with little damping. At frequencies well above the associated resonance frequency the source is in its free state. This has to be the case for all degrees of freedom. The free velocity can then be recorded during source operation for the desired degrees of freedom. The measurement setup is shown in Fig. 5.4(a). The washing machine is suspended on four springs and the ideal source on four rubber cords. The spectrum of the free velocity at foot 3 of the washing machine is shown in Fig. 5.6 together with the reproducibility at all feet. The reproducibility was obtained by repeating the measurement after taking apart the complete setup. Before repeating the measurement the source was run at least once in a different state (blocked or coupled). The frequency range is extended to 10 Hz to show the frequency run-up of the excitation frequency. The predictions only consider frequencies above 22 Hz.

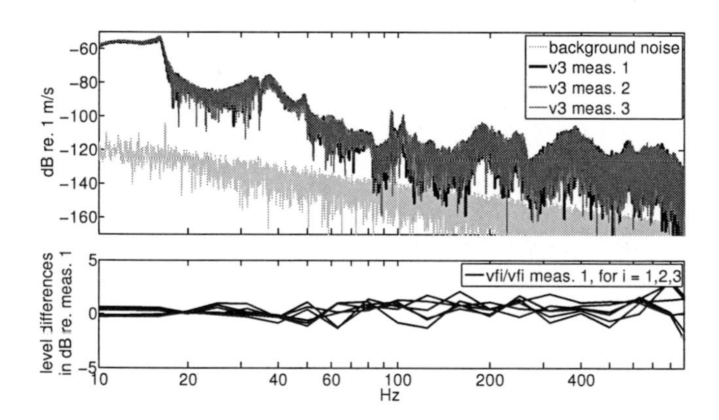

Figure 5.6: Free velocity measurement of the washing machine and the measured reproducibility.

5.3.2 Blocked force

The blocked force is measured on a 20 cm concrete floor. It was verified with a measurement that the mobility of the floor is 10 times lower than the source mobility. The sources with their connected force transducers are attached to the immobile base through a steel disc that is glued to the floor with dual compound epoxy glue. This is depicted in Fig. 5.4(c). The normal blocked force and its reproducibility as it was measured here are shown in Fig. 5.7. The reproducibility was also obtained by repeating the measurement after taking apart the complete setup. Between the measurements the source was run at least once in a different state (free or coupled).

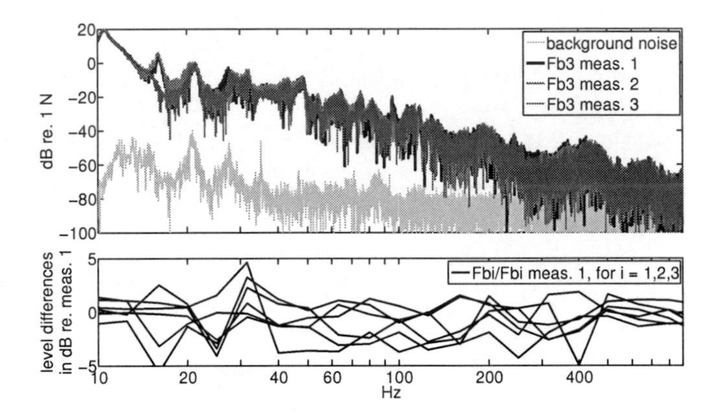

Figure 5.7: Blocked force measurement of the washing machine and the measured reproducibility.

5.3.3 Operational velocity

The operational velocity was measured during run-up at the contact points of the coupled system in the normal direction to the floor. The velocity spectrum and the reproducibility are shown in Fig. 5.8.

5.3.4 Source mobility

The mobility of a source can be measured in its freely suspended state [28]. To obtain the mobility of the washing machine every foot is excited through a force transducer and the velocity is measured simultaneously at all feet in the normal direction. The setup is shown in Fig. 5.4(b). The velocity was measured on-axis, 2.5 cm above the actual foot surface. It was verified with three accelerometers

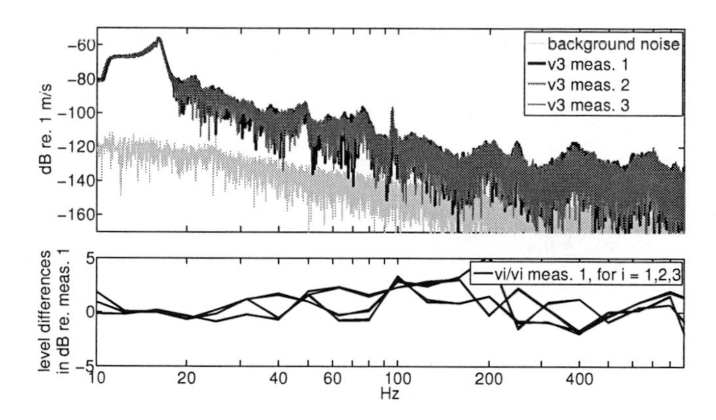

Figure 5.8: Operational velocity of the washing machine and the measured reproducibility.

that measuring at this location gives the same result as the mean of two sensors off-axis but in-plane with the surface of the foot as shown in Fig. 5.9(b).

Three different measurement methods were investigated to obtain the source mobility: a shaker excitation through a 2 mm thick and 60 mm long steel stinger (shaker), an impulse excitation by use of a shaker (automated impulse) and a conventional impulse hammer excitation (impulse hammer). The different setups are shown in Fig. 5.9. The measurement of the source mobility is the most difficult of all independent measurements. The excitation with an impulse hammer is not convenient because the hammer has to be hit upwards. To avoid this an automated hammer was developed. The disadvantage of using a shaker with a stinger is the restriction of the source for the non-normal components. At low frequencies this showed to be negligible. Above 200 Hz the driving point mobilities deviate by more than 10 dB compared to impulse hammer method.

In Fig. 5.10 the point impedance (after matrix inversion) is shown at foot 2. Below 200 Hz the measurements show good agreement whereas differences of more than 10 dB are found above 200 Hz. The differences have to attributed to the restriction of the source through the stinger. It was tried to increase the length of the stinger in order to reduce the influence but this lead to even stronger errors at the resonance frequencies of the longer stinger.

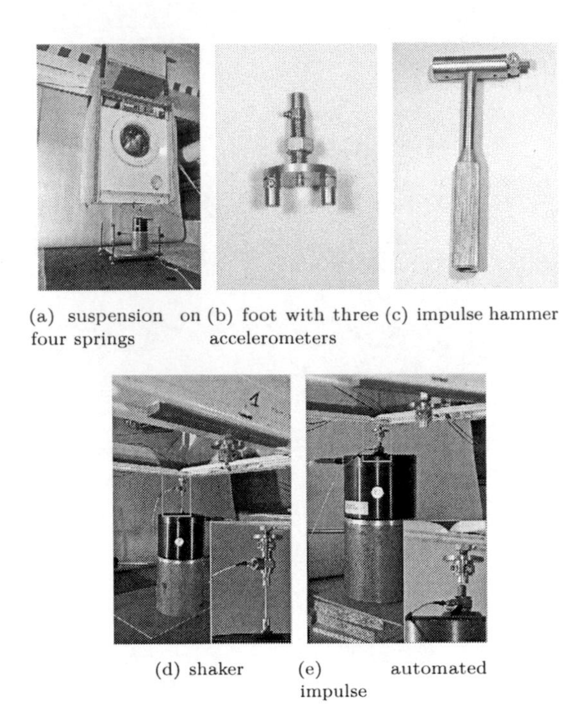

(a) suspension on four springs (b) foot with three accelerometers (c) impulse hammer

(d) shaker (e) automated impulse

Figure 5.9: Setup of three different excitation sources for the source mobility measurements.

The measurements with the automated impulse and the impulse hammer are very similar. The impulse hammer is less convenient to use and the use of averaging to increase the SNR is limited because of poor repeatability. The operator is part of the measurement equipment and this involves considerable influence on the result depending on concentration, posture, technique and experience. However considering those facts, the impulse hammer delivers very good results if the measurements are carefully performed. Because of the similarity to the automated impulse method, the results of the impulse hammer are not used in any of the following analyses for the washing machine.

The accuracy of the mobility measurements can be verified by the use of the

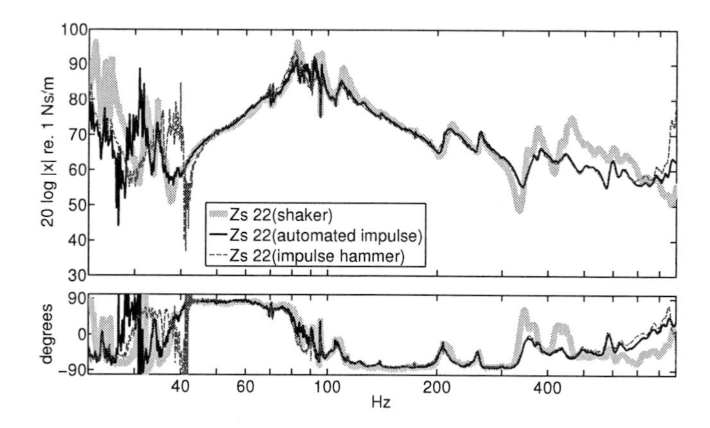

Figure 5.10: Point impedance of the washing machine at foot 2.

property of reciprocity: mobility matrices are symmetric, hence Y_{ij} should be identical to Y_{ji}. This is investigated for both remaining excitation methods in Fig. 5.11. Below 125 Hz the reciprocity of the shaker measurements is clearly superior and almost perfect. Above 125 Hz the automated impulse method is better even though some strong peaks are still present. It shows that the shaker measurements in Fig. 5.10 are wrong at high frequencies. A more general conclusion from Fig. 5.11 is that the measurement of the transfer mobilities involves considerable errors independent of the measurement method.

The repeatability of both measurement methods (not shown) was verified to be below 0.1 dB over the frequency range of interest. The reproducibility of the measurements was obtained by measuring again after taking apart the complete setup. The results for both methods in Fig. 5.12 are equally good. Below 200 Hz deviations up to 6 dB are seen. Above 200 Hz the automated impulse method shows again to be the better choice of the two. It is expected that the error in reproducibility between laboratories will increase even more as it was observed during a round-robin test of mobility measurements in [38]. Especially variations of the measured resonance and anti-resonance frequencies are expected. For a standardisation of source characterisation methods such effects will also be important but are not dealt with here. A detailed investigation into this matter can be found in [4].

The above investigation shows that it is not trivial to precisely determine the source mobility. The overall accuracy is good but not perfect. It is remarkable that it was simply not possible to obtain more consistent results below 50 Hz in Fig. 5.10. The use of a second automated low-frequency impulse hammer with a heavy mass could lead to an improvement of the results.

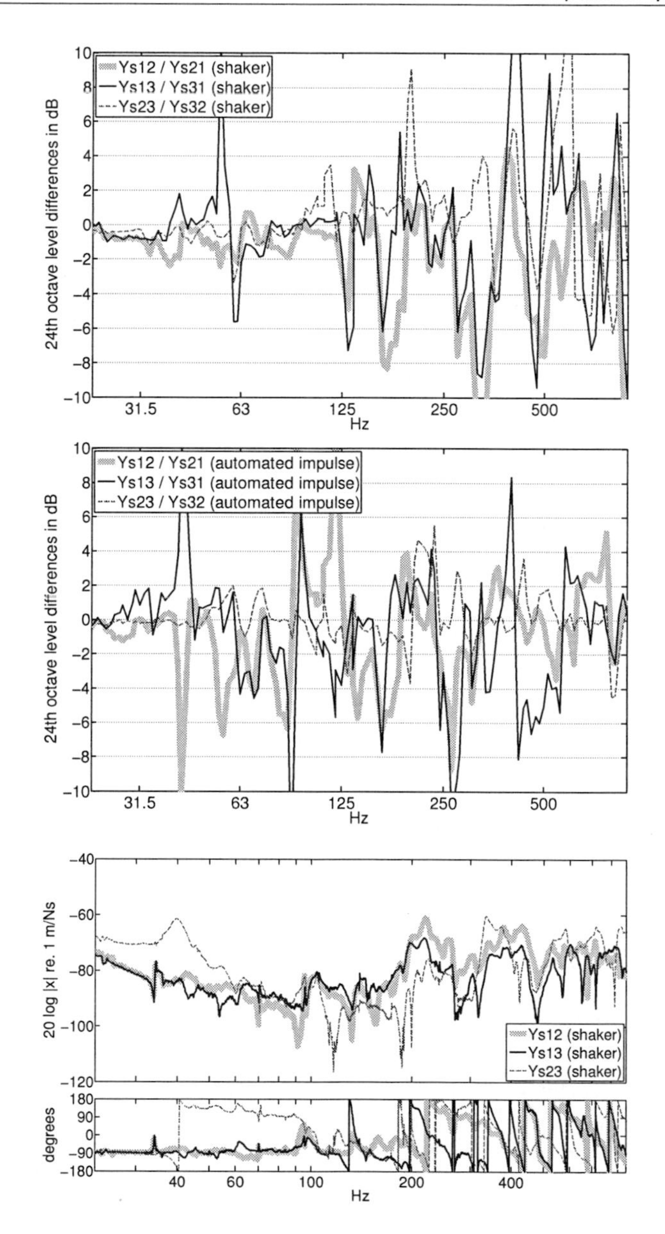

Figure 5.11: Reciprocity analysis of the source mobility of the washing machine.

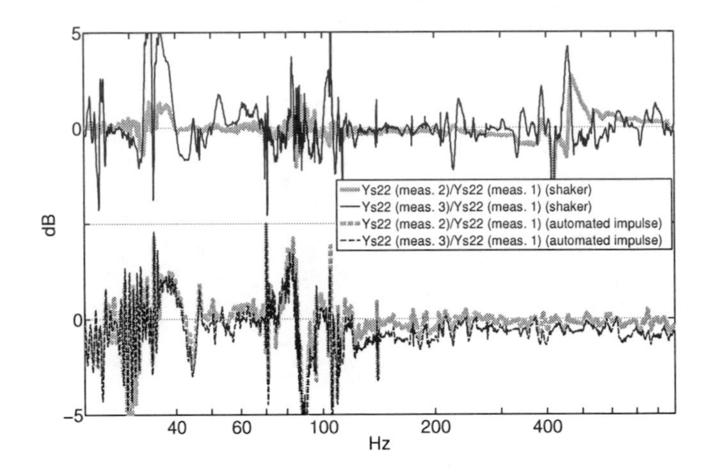

Figure 5.12: Reproducibility of the source mobility of the washing machine.

The source mobility of the ideal source could be measured more accurately because of its simple design. Especially at low frequencies the error in the reciprocity analysis is less compared to the washing machine as shown in Fig. 5.13. Also the difference in source mobility according to the different measurement methods is in the range of ±1 dB.

5.3.5 Receiver mobility and transfer paths

The measurement principles and methods in section 5.3.4 also apply to the measurement of the floor mobility. The floor is excited from above and the acceleration is measured 1.5 cm off-axis from below as shown in Fig. 5.4. The same analyses as in section 5.3.4 were performed for the floor mobility. To reduce the amount of figures only the reciprocity at position 1 is shown in Fig. 5.14. It is clear from those results that the measurement of the floor mobility results in more reliable data independent of the measurement method. The deviation of the reciprocity and the reproducibility spectra are below 2 dB (not shown). Because of a higher SNR the shaker measurements are used throughout the chapter.

The transfer paths from the contact points to the microphone positions were measured simultaneously with the floor mobility. As long as the measurement

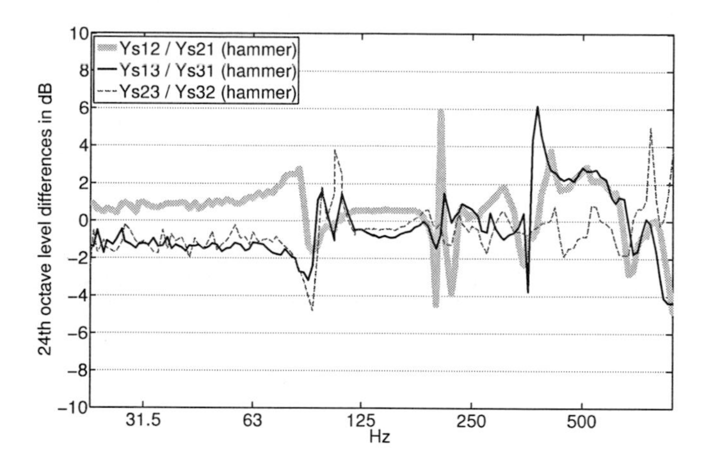

Figure 5.13: Reciprocity analysis of the source mobility of the ideal source.

provide a suffcient SNR, vibro-acoustic measurements are in general less prone to errors and relatively easy to perform.

5.3.6 Coupled mobility and transfer paths

For the measurement of the coupled system the excitation is applied with an impulse hammer from below as close as possible to the contact points. The acceleration is measured on top 1.5 cm away from the excitation point. The reciprocity analysis in Fig. 5.15 shows that the measurement of the coupled transfer mobilities involves considerable errors. The measurements were performed with an impulse hammer to save time. Furthermore, a shaker excitation from below was experimented with. It was however impossible to shield the sound radiation from the shaker to such a degree that the transfer paths could be measured simultaneously. With an impulse hammer this can easily be achieved. In that case the body of the operator is inside the room where the transfer paths are measured. A comparison with a shaker excitation from above showed that the effect of the operator could neglected.

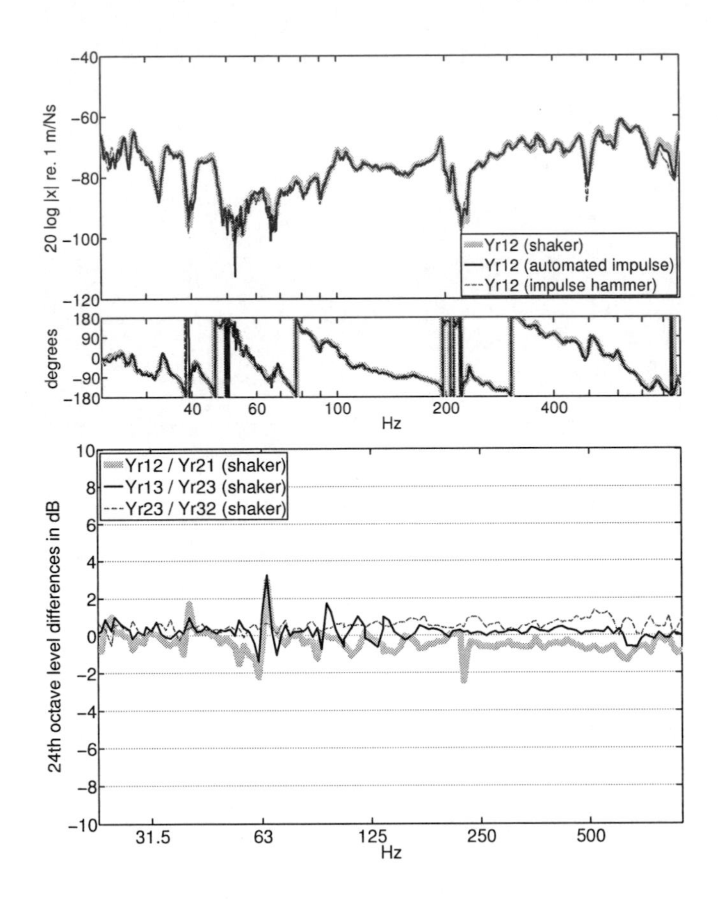

Figure 5.14: Reciprocity analysis of the floor mobility at position 1.

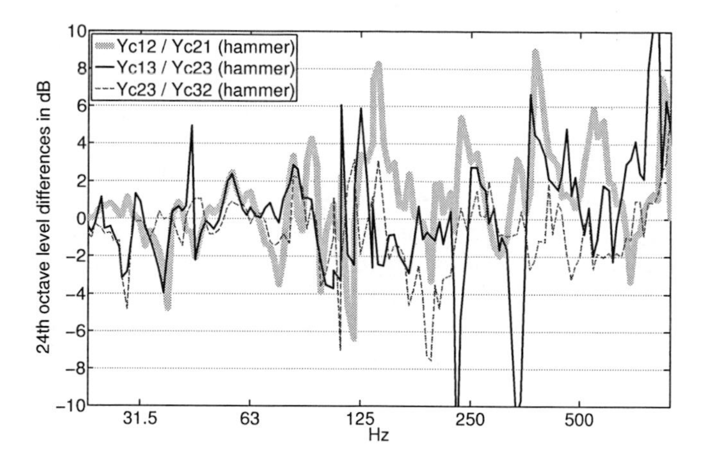

Figure 5.15: Reciprocity analysis of the mobility of the washing machine coupled to the floor at position 1. Measured with an impulse hammer from below.

5.4 Results

The sound pressure in the receiving room is predicted based on measurements of the source mobility, receiver mobility, blocked force, free velocity and the transfer paths. All measurements only consider the normal components. The likelihood of moments is reduced in the setup by the small contact area between the source and the receiver. The remaining degrees of freedom are neglected.

The results will be presented as ratios of the predicted sound pressure in third octave bands over the directly measured sound pressure in third octaves, known as sound pressure level differences. The analyses are limited to third octave resolution because of psychoacoustic reasons. The frequency range of the analyses is 22 to 890 Hz as it was previously discussed. The average over the microphone positions is the average of the band *levels* of the ratios. Third octave bands with an SNR below 10 dB are not taken into account. The division of the third octave band sound pressure levels is a non-linear operation. As a result the sound pressure ratios will depend to some extent on the spectral content of the excitation signal. A more precise and linear approach would be to divide the narrow band spectra directly and calculate third octave bands afterwards. The disadvantage of this more precise method is the fact that very small shifts between peak values of the calculated and directly measured sound pressure lead to very large values of the ratio. The psychoacoustic relevance of a small shift of those peak values is not important and usually inaudible. Therefore the third octave bands are calculated before the division. For the evaluation of the results the non-linear effect inherent to this kind of analysis has to be borne in mind.

5.4.1 Quantities

The most important investigation is the prediction according to Eq. 2.4 and 2.5. The predicted sound pressure depends on the independent measurement of the source activity and on the calculated coupling. To investigate the influence separately a few new quantities based on coupled measurements are introduced below. The averaging process described above is not explicitly shown in the legends of the figures and in the quantities. The predicted averages will be referred to as ratios denoted by $p_{\text{predicted}}/p_{\text{directly measured}}$. All results are evaluated from five microphone positions.

From Eq. 2.8 and 2.4 the sound pressure is calculated as

$$p\text{–}ysyr\text{–}fb = \mathbf{H}(\mathbf{Y}_s + \mathbf{Y}_r)^{-1}\mathbf{Y}_s\mathbf{F}_b \tag{5.1}$$

Errors result from the calculation of the coupling due to errors in source and receiver mobility and from errors in the measurement of the blocked force. The influence caused by errors of the transfer paths is insignificant compared to the errors at the vibrational level.

Eq. 2.8 and 2.5 yield

$$p\text{–}ysyr\text{–}vf = \mathbf{H}(\mathbf{Y}_s + \mathbf{Y}_r)^{-1}\mathbf{v}_f \tag{5.2}$$

Errors again result from errors in the source and receiver mobility and from errors in the measurement of the free velocity.

From Eq. 2.20 and the directly measured blocked force, the sound pressure is calculated as

$$pfb = \mathbf{H}_c\mathbf{F}_b \tag{5.3}$$

This prediction separates the directly measured blocked force from the coupling. The coupling is included in the coupled transfer paths.

Eq. 2.20 and 2.19 yield

$$pfbc = \mathbf{H}_c\mathbf{F}_{bc} \tag{5.4}$$

This prediction is entirely based on measurements in the coupled state.

Eq. 2.8 and 2.19 yield

$$p\text{–}ysyr\text{–}fbc = \mathbf{H}(\mathbf{Y}_s + \mathbf{Y}_r)^{-1}\mathbf{Y}_s\mathbf{F}_{bc} \tag{5.5}$$

After verifying $\mathbf{F}_{bc}$ with Eq. 5.4 this equation reveals the influence of the calculation of the coupling separately.

The high-mobility source case is calculated from the coupled blocked force to present the best approximation independent of errors in the directly measured blocked force. Eq. 2.8, 2.6 and 2.19 yield

$$p\text{–}highys \approx \mathbf{H}\mathbf{F}_{bc} \tag{5.6}$$

Eq. 2.8 and 2.7 yield

$$p\text{–}lowys \approx \mathbf{H}\mathbf{Y}_r^{-1}\mathbf{v}_f \tag{5.7}$$

5.4.2 Mobility comparison

Before looking at the prediction it is important to know the coupling situation at the different positions on the floor (Fig. 5.3). The mobility of the washing machine in comparison to the floor mobility at the different positions is shown in Fig. 5.16. At all positions strong coupling is observed. The source never really exhibits clear high or low-mobility source characteristics. At low frequencies very high coupling exists at particular frequencies where identical magnitudes of the source and the receiver mobilities are found in conjunction with an opposite phase.

The positions on the floor were selected to provide two similar locations in terms of coupling with the source. Position 2 is only slightly shifted parallel to the joists from position 1. The contact points for both positions are at the same distance from the joists and present a very comparable coupling situation. The third location differs more from position 1 and 2. At this position, contact points 2 and 3 are in the middle of the bay between the joists. The mobility at those points is much higher compared to contact point 1. The different positions on the floor were selected to present a variety of coupling situations for the washing machine.

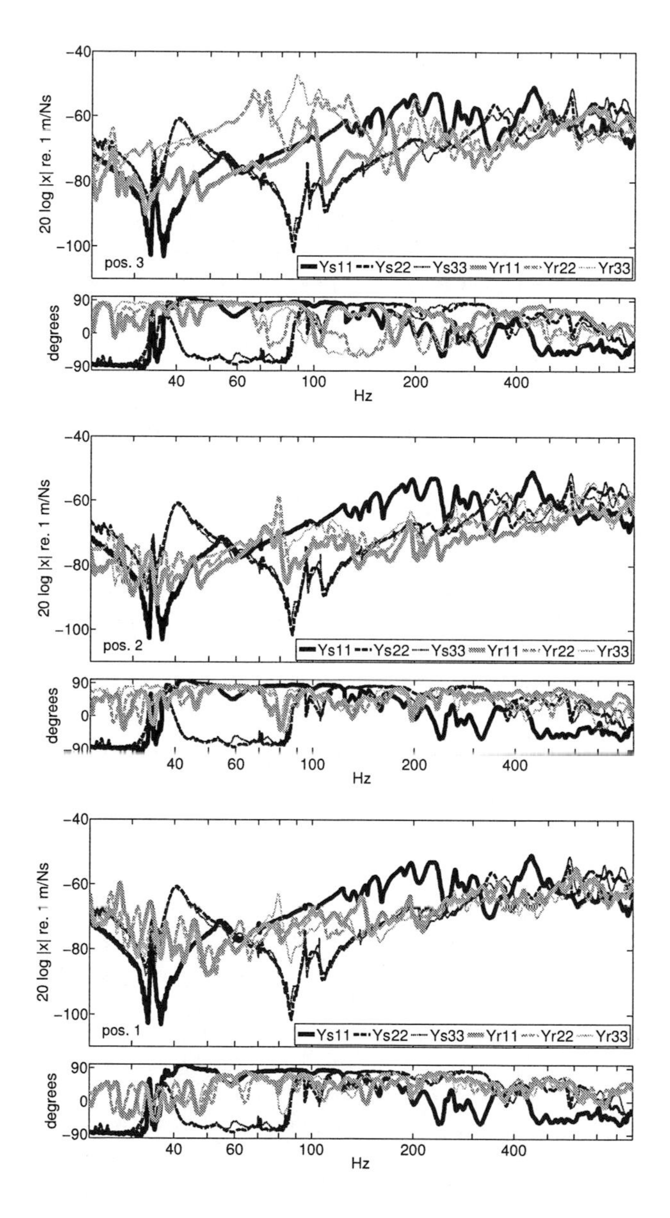

Figure 5.16: Mobility of the washing machine in comparison to the mobility of the three positions on the floor (Fig. 5.3). From bottom to top figure: position 1,2,3.

5.4.3 Predictions

The prediction of *p–ysyr–fb* and *p–ysyr–vf* for the three different locations on the floor is shown in Fig. 5.17. The results are calculated by use of the source mobility measured with a shaker and the source mobility measured with the automated impulse excitation. The differences according to both measurement methods are very small. Above 250 Hz the impulse results are closer to 0 dB than the shaker results. The reciprocity of the automated impulse measurements was found to be better than the shaker measurements which corresponds to the improved prediction found here. At low frequencies no difference in the prediction is seen for the different measurement methods although the reciprocity of the shaker measurements was found to be better. Because of the small differences between both methods and to reduce the amount curves the following results will be based on a source mobility that consists of the shaker measurements below 125 Hz and automated impulse measurements above 125 Hz.

Fig. 5.17 is the most important result of this chapter and shows that the prediction out of the independent quantities leads to reasonable results for the washing machine on the different locations on the floor. The prediction works quite well above 125 Hz with deviations between ±5 dB. Below 125 Hz unacceptable differences of more than 20 dB are predicted. The largest errors are found in *p–ysyr–fb*.

5.4.3.1 Reproducibility

The reproducibility of the measured source activity, mobility and transfer paths was documented in section 5.3. Those reproduced measurements can be used to obtain the reproducibility of the prediction. For two reproduced measurements, the sound pressure ratio depends on five independent variables and allows for 32 combinations to be calculated. This is shown in Fig. 5.18 for position 1 and applies to all of the following results for the washing machine. The curves clearly highlight two sets of data within *p–ysyr–fb* and *p–ysyr–vf* attributed to the low reproducibility of the source activity, c.f. Fig. 5.6 and Fig. 5.7. Below 200 Hz the spread of 5 dB is a considerable variation that has to be kept in mind. It should be pointed out again that a reproduced measurement of the blocked force was preceded by a measurement of the free velocity to make sure that the worst case is demonstrated. Because of the effort involved in repeating

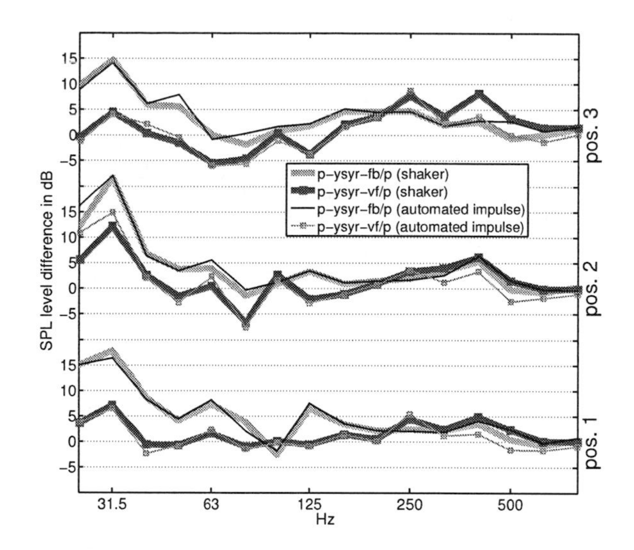

Figure 5.17: Predictions for all locations of the washing machine on the floor. The source mobility is measured with the shaker (shaker) and an impulse excitation of the shaker (automated impulse).

those measurements, not more than two reproduced measurements were made in this way. Ideally a source characterisation should involve more reproduced measurements in combination with different spectral content of the excitation to improve the estimate of the standard deviation and to allow for an analysis of the significance of different results. A higher amount of measurements would also average out the influence of the non-linear behaviour of third octave band analyses on the prediction.

The influence of the mobilities and the transfer paths is much lower. It is illustrated with a separate calculation of 8 combinations in Fig. 5.18.

5.4.3.2 Matrix inversion

The 2-norm condition number of a matrix is a commonly used measure to indicate how well-conditioned a matrix is. If a measured or calculated matrix

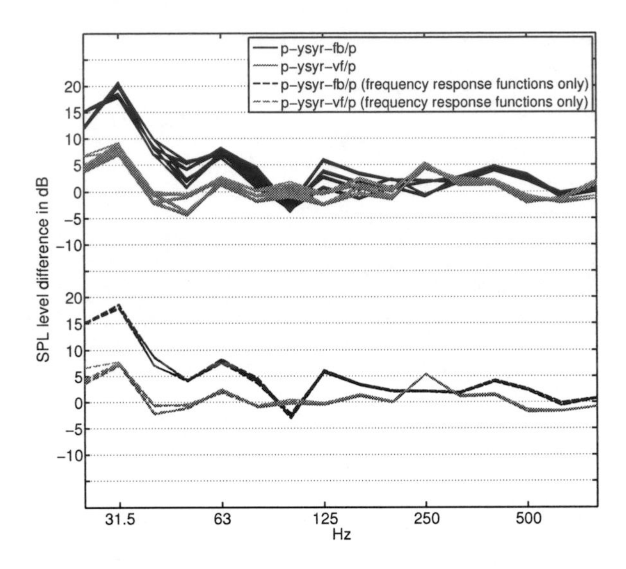

Figure 5.18: Reproducibility of the prediction at source position 1 estimated from a calculation of 32 combinations based on two reproduced measurements.

needs to be inverted, the condition number must be examined to verify the result of the inversion. Condition numbers close to 1 allow for the matrix to be inverted without producing large errors. Information about the error due to the inversion depending on the value of the condition number is only difficult to obtain [39]. It should also be pointed out that the condition number does not contain information about the accuracy of the measured data. A perfectly measured mobility matrix might result in a highly inaccurate impedance matrix simply due to the relationship between the (perfect) transfer mobilities.

The calculated interaction forces are based on the inversion of the sum of the source and receiver mobility according to Eq. 2.4 and 2.5. All inversions involved in the calculation of the quantities in section 5.4.1 are shown in Fig. 5.19. Above 125 Hz the condition numbers are below 10. At low frequencies numbers up to 100 are observed at a few discrete frequencies. Nevertheless the condition numbers show that it is viable to calculate the interaction force and the coupled mobility from the measured data.

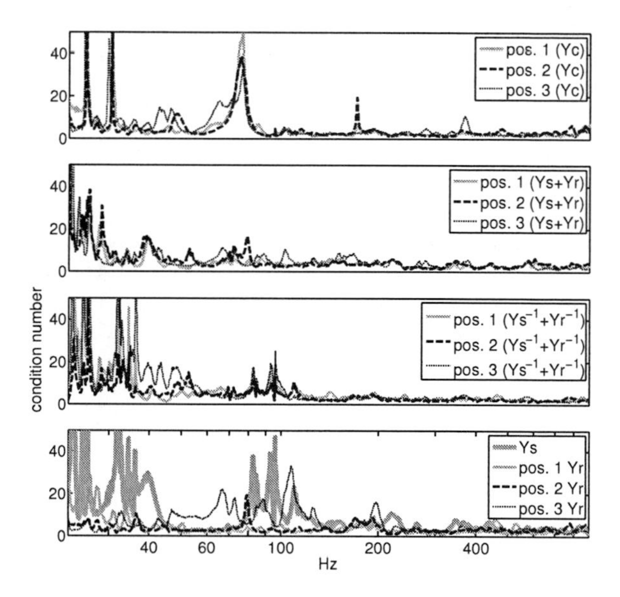

Figure 5.19: Condition number of the mobility ratios in Eq. 2.4 and 2.5.

The general results for the washing machine on the wooden floor in Fig. 5.17 will now be analysed in more detail and compared to predictions based on different (coupled) quantities to clarify the deviations at low frequencies. Finally the results will be compared to predictions based on certain assumptions to find out whether it is allowed to further simplify things in practice.

5.4.3.3 Blocked force

The predicted sound pressure based on the blocked force showed to overestimate the sound pressure at low frequencies in Fig. 5.17. The sound pressure ratios p–$ysyr$–fb are shown separately for the different positions on the floor in Fig. 5.20. To separate the influence of errors in the blocked force from errors in the calculated coupling, the blocked force can be obtained in the coupled state according to Eq. 2.19. The coupled blocked force in Fig. 5.20 produces almost perfect predictions of the sound pressure in $pfbc$. In terms of accuracy for applications in building acoustics the results are very good. This aspect was the topic of investigation in [27].

The coupled blocked force is now used for the prediction of the interaction force in Eq. 2.4 and reveals the influence of errors due to the calculated coupling only. This prediction, denoted by *p–ysyr–fbc*, involves large errors at 32 Hz and proves that the calculation of the coupling fails at low frequencies. However, the directly measured blocked force is also responsible to some extent for the result of *p–ysyr–fb*. This is investigated by looking at *pfb* which is calculated from the coupled transfer paths and the directly measured blocked force. Also those curves increase at low frequencies in Fig. 5.20. Due to the non-linearity of the third octave band analysis the errors in *pfb* and *p–ysyr–fbc* do not add up to *p–ysyr–fb* but it is shown that both, the directly measured blocked force and the the calculation of the coupling are responsible for errors in the prediction of *p–ysyr–fb*.

The reason for the errors in the directly measured blocked force have been discussed in section 5.3.2. The calculation of the coupling will be analysed in more detail in section 5.4.3.8.

5.4.3.4 Low and high-mobility sources

The assumption of an ideal high-mobility and low-mobility source are frequently used in practical source characterisation. The assumption simplifies the calculation of a prediction and therefore reduces the possibility of errors in the calculation. However, if coupling between the source and the receiver exists the assumption cannot be applied. In the mobility comparison for the washing machine in Fig. 5.16 no ideal source behaviour is found. This is supported by the predictions shown in Fig. 5.21. The results in Fig. 5.21 clearly show that it is worthwhile to account for the coupling. The ideal cases introduce errors up to 15 dB over broad frequency ranges. It is interesting that combinations of *p–lowys* and *p–highys* would be very good estimates over most of the frequency range. However by only investigating the mobilities in Fig. 5.16 it is not possible to determine where to use which ideal case.

Fig. 5.21 also shows the prediction out of the free velocity in comparison to the blocked force and the coupled blocked force. Especially the difference between

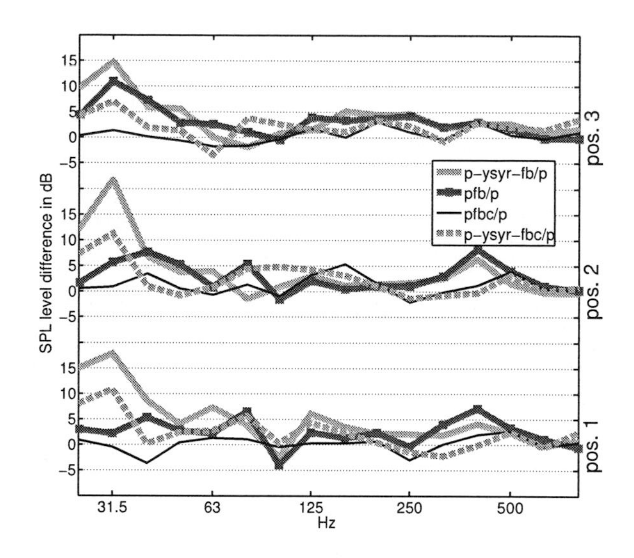

Figure 5.20: Prediction of the sound pressure out of the source and receiver mobilities and the blocked force.

the predictions out of the free velocity and the coupled blocked force are surprisingly low if one considers the different states the measurements are based on. The coupled blocked force leads to better predictions compared to the directly measured blocked force. This will be investigated in more detail in section 5.4.3.7.

5.4.3.5 Transfer mobilities

The importance of the transfer mobilities is illustrated at a single position in Fig. 5.22. Below 125 Hz the interaction between the contact points cannot be neglected. At higher frequencies the transfer mobilities are less dominant but still within 10 dB from the point mobilities. In Fig. 5.22 the maximum level always depends on one of the point mobilities which again emphasizes their importance.

To evaluate the importance of the interaction between the contact points on

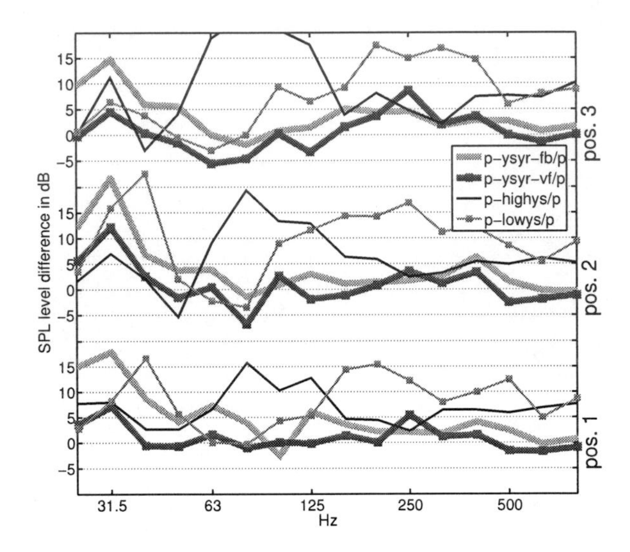

Figure 5.21: Predictions in comparison with the high-mobility and low-mobility cases.

the prediction of the sound pressure, the prediction is compared to a calculation without interaction. This is achieved by omitting the transfer mobilities. *p–ysyr–diag–fb* and *p–ysyr–diag–vf* denote the quantities without transfer mobilities (diagonal matrix). It should be pointed out that the calculation and the measurement of the transfer mobilities does not impose a significant additional effort. For sources with three contact points it involves the simultaneous measurement with two additional accelerometers. However, the accuracy of the prediction might be influenced by errors due to the lower SNR of the transfer mobilities.

The differences between the calculation with and without transfer mobilities are shown in Fig. 5.23. For all positions on the floor the differences are negligible above 100 Hz. At low frequencies there is on average no advantage of considering the coupling between the points. Ignoring the interaction happens to result in a better prediction for this case study. Section 5.4.3.8 will elaborate this aspect.

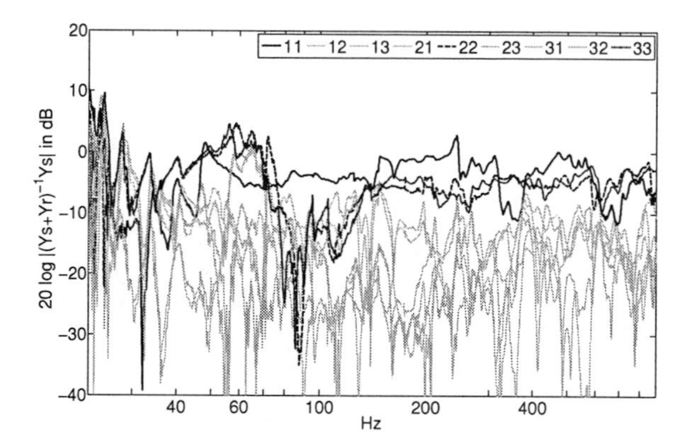

Figure 5.22: Calculated factor $(\mathbf{Y_s} + \mathbf{Y_r})^{-1}\mathbf{Y_s}$ (Eq. 2.4) out of the measured source and receiver mobility for position 1 on the floor.

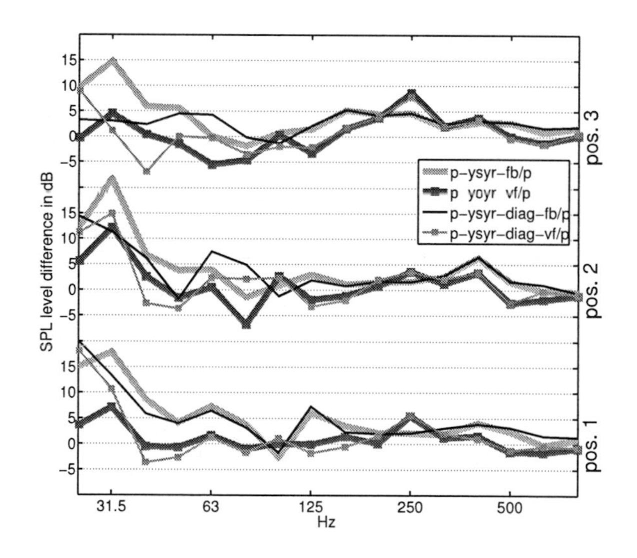

Figure 5.23: Comparison of the predictions with and without interaction between the contact points. No interaction is denoted by *diag*.

5.4.3.6 Phase between the contact points

The calculation of the coupling between the source and receiver according to Eq. 2.5 relies on accurate information about the phase of the source and the receiver. The sum of source and receiver mobility in the denominator can, in theory, lead to very high interaction forces. This is likely to occur only at low frequencies where source and receiver exhibit mass and spring behaviour over a broad frequency range.

The reception plate method [19] used for source characterisation does not yield the phase of the source mobility. The phase between the contact points has to be assumed to be zero or random. The effect of this assumption on the prediction is investigated in Fig. 5.24. The phase of the source and receiver mobilities was chosen to be zero or random. As expected, strong errors are introduced but the assumption of a random phase leads to a reasonably accurate prediction. Above 125 Hz the errors are on average below 5 dB.

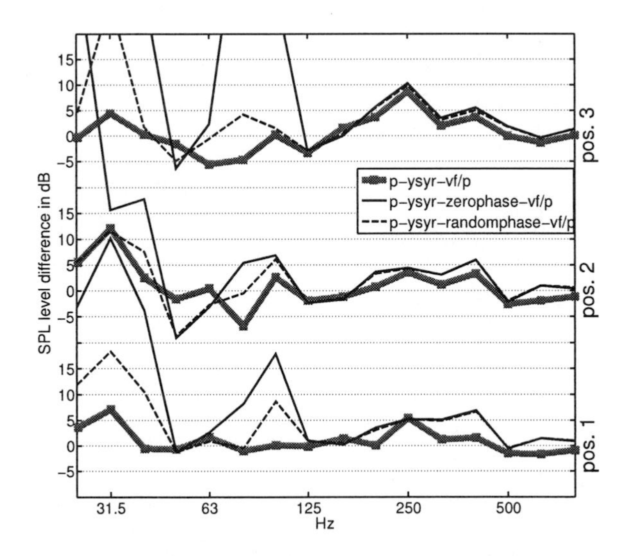

Figure 5.24: Comparison of the predictions with a zero phase and random phase assumption of the source and receiver mobilities.

5.4.3.7 Coupled blocked force

It was shown in Fig. 5.20 that the coupled blocked force better represents the source activity compared to the directly measured blocked force. The coupled blocked force is calculated from measurements in the coupled state at a specific position on the floor. The fact that an independent quantity is obtained in the coupled state is a very strong contradiction. If it is to be used in practice the coupled blocked force determined in one coupling state (at one position) should at least be very similar to the coupled blocked force determined in a similar state of coupling. Measuring the coupled blocked force for every situation that has to be predicted has of course little value. Instead it might be valuable to obtain the coupled blocked force for a certain class of source and receiver structures and develop a method that leads to a more precise prediction of the coupling within this specific class.

If for the case study at hand, the class is defined as a washing machine on a bare wooden floor, the coupled blocked force should not vary depending on the position on the floor. The comparison is shown in Fig 5.25. Below 100 Hz the results for coupled blocked forces are within ±10 dB and prove to be better than the results for the directly measured blocked force. Above this frequency there is no advantage to using the coupled blocked force.

5.4.3.8 Coupling

Considering the effort taken in obtaining accurate source and receiver mobilities and (reproducible) source activity quantities, the predictions in Fig. 5.17 and 5.20 are not satisfactory. The predictions depend on the calculation of the coupling and on the source activity term. The use of the coupled blocked force should however result in a perfect prediction if the coupling is correctly calculated. Nevertheless deviations up to 10 dB are observed in *p–ysyr–fbc*. To investigate the calculation of the coupling in more detail the coupled mobility according to Eq. 2.2 can be compared to the directly measured coupled mobility. This analysis is shown for foot 2 in Fig. 5.26. Above 200 Hz the differences between the measured and calculated coupled mobility are below 5 dB. At low frequencies however large differences are found that have to be attributed to the failing degrees of freedom in the calculation. The likelihood of moments was reduced by choosing contact points with a very small surface of 9 mm in diameter. The small

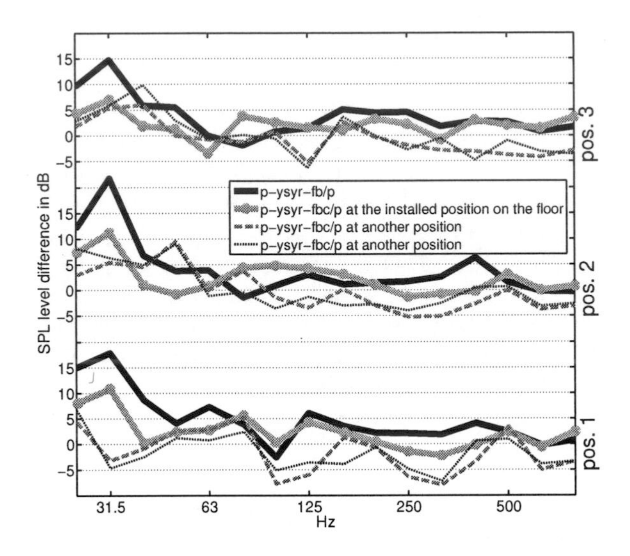

Figure 5.25: The coupled blocked force obtained at the different positions on the floor and used in Eq. 2.4.

steel cylinders on top of a relatively soft chipboard plate are very unlikely to result in strong moments below 200 Hz. The conclusion from those observations is that the translational x and z-components cannot be neglected for a correct calculation of the coupled mobility. However below 50 Hz the predictions might also suffer from the high condition number in Fig. 5.19. To rule out this uncertainty the measurements were repeated with an idealised source. This will be addressed in the next section.

In Fig 5.23, the calculations without interaction between the contact points were better at low frequencies compared to the calculations with interaction. Fig. 5.26 proves that the contradictory findings are not related to the calculation of the coupling between source and receiver. The omission of the interaction (*diag*) just leads to different and less accurate predictions that happened to compensate for errors in the blocked force in *p–ysyr–diag–fb*. Fig. 5.26 clearly shows that the calculation of the coupled mobility does improve if the transfer mobilities are taken into account.

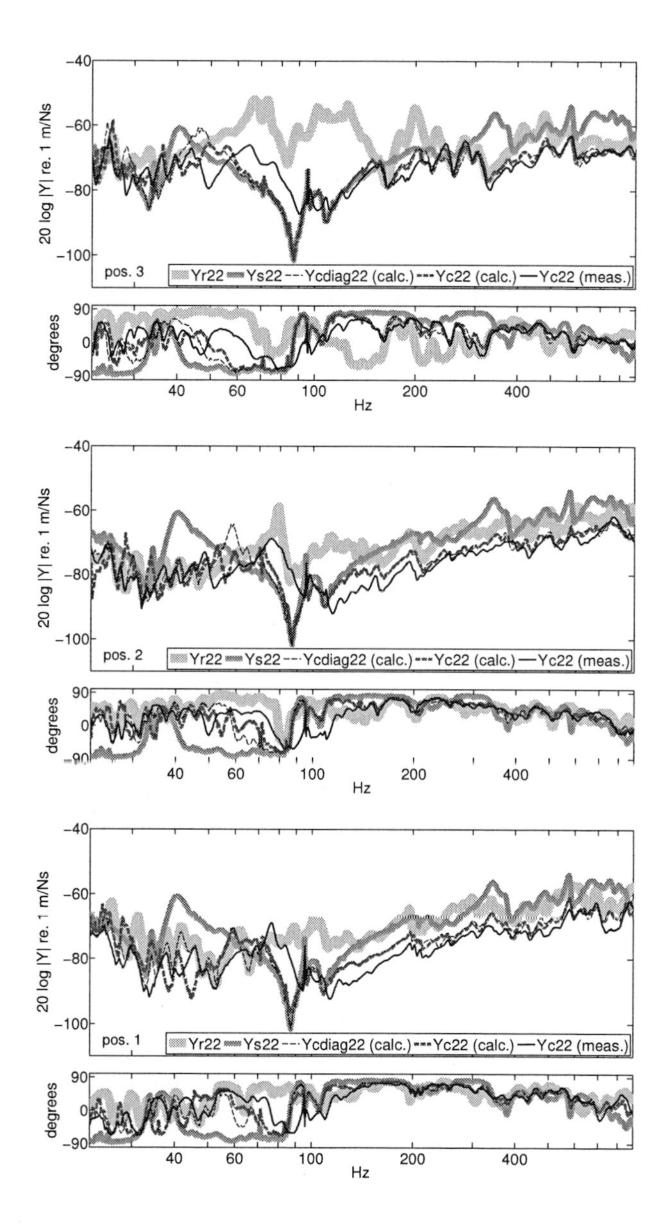

Figure 5.26: Comparison of calculated (c.f. Eq. 2.2) and measured coupled mobilities of the washing machine at foot 2. From bottom to top figure: position 1,2,3.

5.4.3.9 Ideal source

In section 5.3.4 it was shown that the measurement of the washing machine is questionable at frequencies below 50 Hz (see Fig. 5.10). To circumvent this fact the ideal source shown in Fig. 5.4 was used to repeat the measurements. Its mobility as shown in Fig. 5.27 is comparable to the mobility of the washing machine and results in a similar degree of coupling with the floor. The coupling between the feet is shown in Fig. 5.28. The reciprocity analysis of the source mobility in Fig. 5.13 proves that the measured mobility is more accurate compared to the washing machine. The corresponding condition numbers are shown in Fig. 5.29.

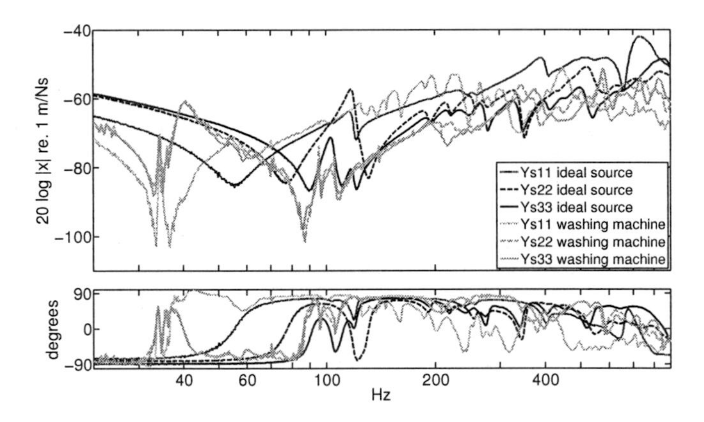

Figure 5.27: Mobility of the ideal source depicted in Fig. 5.4 in comparison to the mobility of the washing machine.

The difference between the coupled mobility directly measured and calculated is shown for foot 3 in Fig. 5.30. The measurements correspond well to the calculations but at crucial frequencies where the source dominates the coupled mobility the differences are very large. The frequency of the first anti-resonance of the source mobility is always shifted upwards in the calculation. The moments were reduced even more for the ideal source by taking feet with a radius of 4 mm. The steel cones accommodate a bolt of 3 mm which leaves a rim of 0.5 mm for the source to stand on the floor. This makes the excitation of moments virtually impossible. The bolt inside the steel cone has a very limited strength

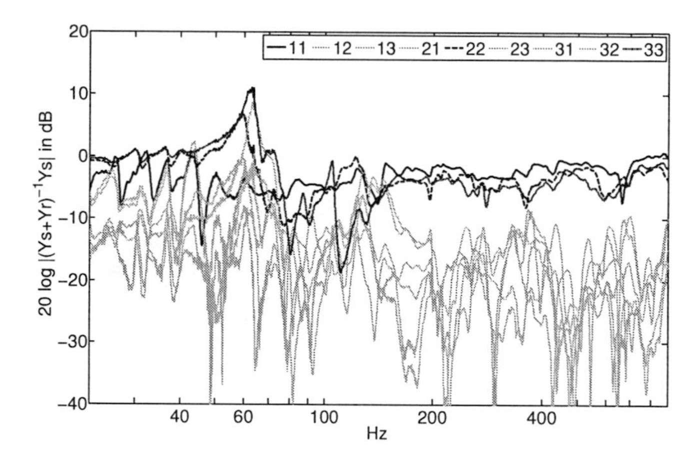

Figure 5.28: Calculated factor $(\mathbf{Y_s} + \mathbf{Y_r})^{-1}\mathbf{Y_s}$ (Eq. 2.4) out of the measured source and receiver mobility for position 1 on the floor.

and primarily fixes the foot in the translational y-direction. From the findings in Fig. 5.30 it has to be concluded again that the translational x and z-components cannot be neglected for a precise calculation of the coupling at low frequencies for the case study at hand. This behaviour was also observed between two firmly fixed beams in [30] as mentioned in [40].

To support the conclusion, a further measurement was performed with a rubber interlayer between the ideal source and the wooden floor at position 2. The diameter of the feet was changed to 17 mm and a rubber layer was simply inserted (without glue) at the three feet. The rubber layer is expected to reduce the coupling through the shear forces compared to the previously used fixed mounting conditions. The source mobility was not measured again but it is assumed that the rubber layer will have very little influence on the source mobility at low frequencies. Hence the calculation of the coupled mobility is still valid at low frequencies. The result is shown in Fig. 5.30 for position 2 on the floor. Around 60 Hz the measurement with the rubber interlayer clearly corresponds better to the calculated mobility. However there are still large differences between the calculation and the measurement that have to be addressed. Whether those differences are due to the remaining coupling in the translational x and z-direction through the rubber would have to be answered by setting up a case study that

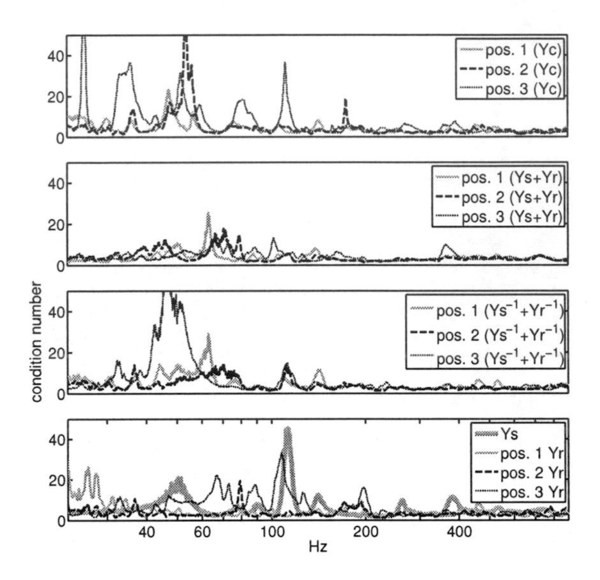

Figure 5.29: Condition number of the mobilities involved in the calculation of the predictions of the ideal source.

incorporates all degrees of freedom in detail. In this work it was chosen to use firmly connected feet to avoid non-linearities and load-dependent characteristics of resilient layers. The frequency dependent measurement of resilient layers is a topic on its own [41, 42] and would have to be accounted for in further case studies. Many sources are resiliently decoupled in practice as it is the most common measure to reduce high-frequency structure-borne sound excitation.

Sound pressure predictions based on the blocked force and the free velocity were also calculated for the ideal source and compared to the direct sound pressure measurements. This presents a prediction of a second source at the same positions on the floor and can be compared to the results of the washing machine in 5.17. The ideal source was run up from 15 to 22 Hz with an out-of-balance mass of 120 g. The results in Fig. 5.31 are similar but less accurate compared to the results of the washing machine. Below 125 Hz the errors are more than 20 dB. This is again an indication that it is not possible to accurately predict the coupling based on translational y-components only. At higher frequencies the curves are on average between ±5 dB.

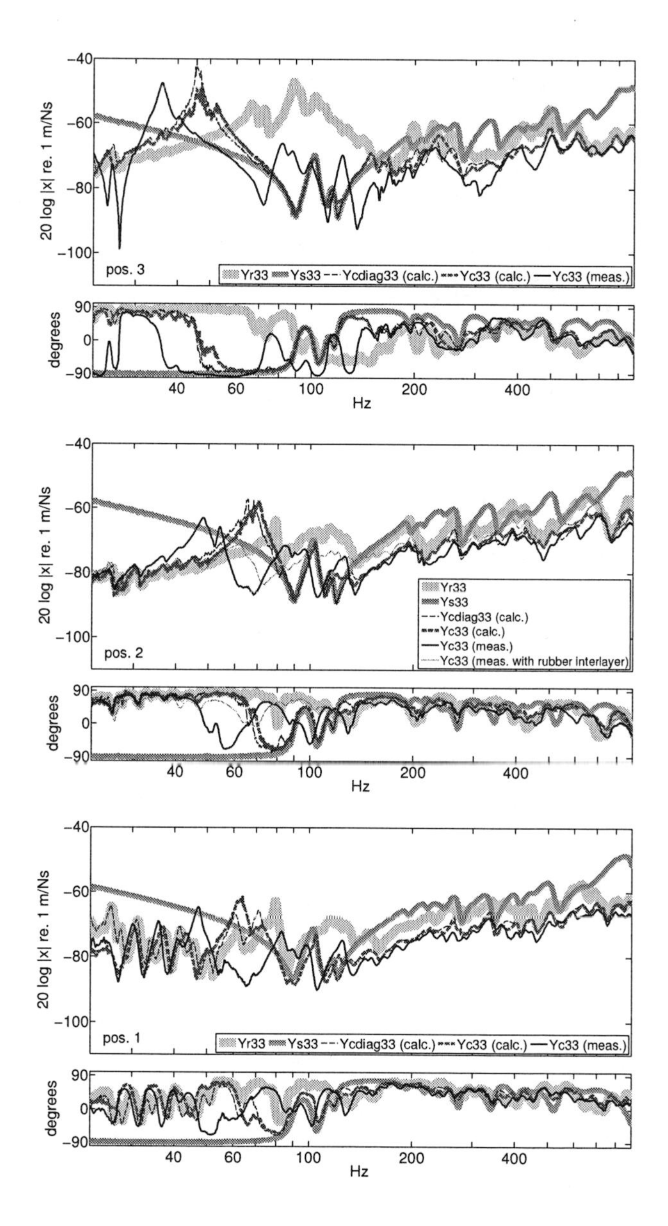

Figure 5.30: Comparison of calculated and measured coupled mobilities of the ideal source at foot 3. Calculated according to Eq. 2.2. From bottom to top figure: position 1,2,3.

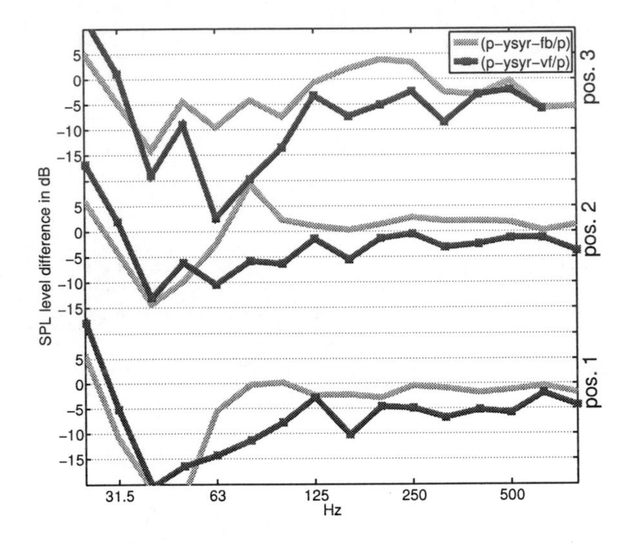

Figure 5.31: Predictions for all locations of the ideal source on the floor.

5.5 Conclusion

A case study was performed with a washing machine on a wooden joist floor. The aim was to evaluate various approaches for an independent source and receiver structure characterisation. The area of the contact points is as small as possible to limit moment excitation. It was shown that the normal components are not sufficient to precisely predict the coupling between the source and the floor below 125 Hz. Above this frequency the predicted sound pressure from the predicted interaction force is within ±5 dB of the directly measured sound pressure. High and low-mobility cases clarified the necessity to account for the interaction and highlight the value of the results. It was shown with measurements on an idealised source that the in-plane shear forces are most likely to be responsible for the errors at low frequencies. In contrast to this conclusion it was shown that the sound pressure could be predicted much more accurately (±3 dB) from the coupled normal components. This means that the normal components are most important for the sound radiation but they are not sufficient to calculate the coupling between a source and a receiver.

The measurement of the source mobility is the most difficult of all independent measurements. Differences of 10 dB were observed between the measurement with a shaker and an impulse hammer. At low frequencies the shaker is better in terms of SNR but in general it can be concluded that the impulse hammer should be used to measure mobilities. To improve the repeatability an automated impulse hammer might be useful in some setups. More advanced techniques of obtaining the source mobility in the coupled state should be investigated [40].

The use of the coupled blocked force showed to be a good alternative to the directly measured blocked force even if it is obtained at a different location of the source on the floor. However, the conclusion about the importance of the shear forces leaves room for improvement for the determination of the coupled blocked force. To correctly obtain the coupled blocked force all relevant degrees of freedom also have to be taken into account.

In practice most sources are resiliently decoupled from the receiver structure and not so often firmly fixed. First measurement results showed that in the case of resilient supports the importance of the in-plane shear forces decreases but remains present. Significant differences of up to 20 dB between the directly measured coupled mobility and the calculated coupled mobility are observed.

This shows once more that it is important to account for all degrees of freedom for the precise characterisation of structure-borne sound sources. The future challenge for source characterisation will be to determine the relevant degrees of freedom for different types of sources. Simplified engineering methods based on that information should then be stated with an expected level of accuracy.

The case study deals with a specific type of washing machine on a bare wooden joist floor. The conclusions are likely to apply to sources with similar dimensions, excitation levels and contact situations. However, it is important to carry out more case studies for different types of source-receiver combinations (sources mounted on walls, gypsum-board walls, sources on rubber mounts, sources with four feet, ...). By doing this different groups of sources can be classified depending on their relevant degrees of freedom.

6 Conclusion and Outlook

To achieve an accurate characterisation of structure-borne sound sources, a tremendous amount of effort is required and to date there is no procedure that leads to precise predictions of sources being coupled to receiver structures. For example highly accurate measurements in NVH applications [43] that take into account all degrees of freedom showed that it is not possible to trace the anomalies to their origin. The inclusion of the rotational degrees of freedom lead to a deterioration of the results in the prediction process. In the field of building acoustics high precision is not required. For the standardisation of structure-borne sound sources in EN 12354-5 [13], a simple and practical method is preferable. It is however crucial to know the level of accuracy involved with certain assumptions (reduced set of degrees of freedom, normal components, reception plate measurements, ...). To gain knowledge about the implications of those assumptions it is necessary to start with precise case studies.

A common assumption for the characterisation of service equipment in building acoustics is to account for the normal components only. This is mentioned for example in EN 12354-5 and can be found throughout the literature on source characterisation in building acoustics. This assumption is the main topic of this thesis. Two important conclusion were drawn:

- **The normal components are sufficient to predict the sound radiation in the coupled state.** The sound pressure was predicted from the measured velocity, mobility and transfer paths in the coupled state. A case study of a washing machine connected to a wooden floor was carried out. This was dealt with in chapter 3 and 4. For feet with a rubber layer and a diameter of 17 mm the deviations between the calculated and predicted sound pressure level are on average 2 ± 2 dB above 125 Hz and 0 ± 2 dB below 125 Hz. For larger feet with firm contact, deviations up to 10 dB were observed at particular frequencies below 125 Hz.

- **The normal components are not sufficient to predict the coupling between a source and a receiver structure from independently measured source and receiver quantities.** A washing machine and a wooden joist floor were measured independently and the sound pressure was predicted from those measurements. This was investigated in chapter 5. The results showed that above 125 Hz the error in sound pressure level is within ±5 dB. However, below 125 Hz errors of up to 20 dB were found.

The ideal procedure for a case study would be to measure all degrees of freedom and then investigate several assumptions by means of calculation. To reduce the complexity of measuring all degrees of freedom a different approach was chosen in this thesis. Only the normal components were measured at all contact points. Moment excitation was minimised mechanically by reducing the contact area at the contact points. The in-plane components were neglected. This approach proved to be valuable in chapter 4. In chapter 5 it would have been advantageous to measure all degrees of freedom because the assumption about the in-plane components turned out to be wrong. Future case studies should include all degrees of freedom even if they are intended for the development of practical engineering methods.

The results obtained in chapter 3 and 4 were used as an intermediate step towards the true prediction in chapter 5. The applicability of measurements in the coupled state is only limited in building acoustics. However, in the field of transfer path analysis in NVH applications the first conclusion is important. The sound pressure level can be accurately predicted from measurements in the coupled state. The most important benefit of this method is the fact that the source does not have to be removed from the structure to obtain the frequency response functions. Some parts in a vehicle for example cannot easily be disassembled (e.g. engine or springs on chassis). It should be pointed out that the calculated total sound pressure is theoretically identical for both the coupled and decoupled state whereas the path contributions are not identical.

The results in chapter 5 are valuable in the context of the characterisation of service equipment in buildings as it is integrated in EN 12354-5. It was shown that a prediction based on normal components leads to considerable errors. Future research is required to determine the relevant degrees of freedom for different groups of source-receiver combinations. Once this is known, it is possible to develop different engineering methods according to the classification. The applicability of EN 12354-5 can only be guaranteed if it is supported by information from extensive research.

Bibliography

[1] A. Mayr. *Vibro-acoustic sources in lightweight buildings.* PhD thesis, The University of Liverpool, 2009.

[2] P. Schevenels. *Investigation of the Source-Structure-Sound Interaction in the Framework of the Source Characterisation and Sound Radiation of Vibrating Sources in Buildings.* PhD thesis, K.U. Leuven, 2011.

[3] A. Elliott. *Characterisation of structure borne sound sources in-situ.* PhD thesis, School of Computing, Science and Engineering, University of Salford, 2009.

[4] T. Evans. *Estimation of uncertainty in the structure-borne sound power transmission from a source to a receiver.* PhD thesis, School of Computing, Science and Engineering, University of Salford, 2010.

[5] M. Späh. *Characterisation of structure-borne sound sources in buildings.* PhD thesis, University of Liverpool, 2006.

[6] DIN 4109. Sound insulation in buildings; requirements and testing, 1989.

[7] ISO 10052. Acoustics - Field measurements of airborne and impact sound insulation and of service equipment sound - Survey method, 2010.

[8] ISO 16032. Acoustics - Measurement of sound pressure level from service equipment in buildings - Engineering method, 2004.

[9] EN 12354-1. Building acoustics - Estimation of acoustic performance of building from the performance of elements - Part 1: Airborne sound insulation between rooms.

[10] EN 12354-2. Building acoustics - Estimation of acoustic performance of building from the performance of elements - Part 2: Impact sound insulation between rooms.

[11] EN 12354-3. Building acoustics - Estimation of acoustic performance of building from the performance of elements - Part 3: Airborne sound insulation against outdoor sound.

[12] EN 12354-4. Building acoustics - Estimation of acoustic performance of building from the performance of elements - Part 4: Transmission of indoor sound to the outside.

[13] EN 12354-5. Building acoustics - Estimation of acoustic performance of building from the performance of elements - Part 5: Sounds levels due to service equipment.

[14] EN 12354-6. Building acoustics - Estimation of acoustic performance of building from the performance of elements - Part 6: Soundabsorption in enclosed spaces.

[15] BASTIAN - The Building Acoustics Planning System. www.datakustik.com.

[16] EN 15657-1. Acoustic properties of building elements and of buildings - Laboratory measurement of airborne and structure borne sound from building equipment - Part 1: Simplified cases where the equipment mobilities are much higher than the receiver mobilities, taking whirlpool baths as an example, 2009.

[17] M. Späh and B. Gibbs. Reception plate method for characterisation of structure-borne sound sources in buildings: Assumptions and application. *Applied Acoustics*, 70(2):361–368, February 2009.

[18] M. Späh and B. Gibbs. Reception plate method for characterisation of structure-borne sound sources in buildings: Installed power and sound pressure from laboratory data. *Applied Acoustics*, 70(11-12):1431–1439, December 2009.

[19] B. M. Gibbs, R. Cookson, and N. Qi. Vibration activity and mobility of structure-borne sound sources by a reception plate method. *J. Acoust. Soc. Am.*, 123(6):4199–4209, June 2008.

[20] C. Höller. Investigation into indirect methods to obtain the mobility of a structure. *Fortschritte der Akustik - DAGA*, 2012.

[21] B. A. T. Petersson and B. M. Gibbs. Towards a structure-borne sound source characterization. *Applied Acoustics*, 61(3):325–343, November 2000.

[22] P. Gardonio and M. J. Brennan. On the origins and development of mobility and impedance methods in structural dynamics. *Journal of Sound and Vibration*, 249(3):557–573, January 2002.

[23] L. Cremer, M. Heckl, and B. A. T. Petersson. *Structure-Borne Sound: Structural Vibrations and Sound Radiation at Audio Frequencies.* Springer-Verlag Berlin and Heidelberg GmbH & Co. K, 2004.

[24] F. Fahy and J. Walker, editors. *Advanced applications in acoustics, noise and vibration.* Spon Press, 2004.

[25] S. Rubin. Transmission matrices for vibration and their relation to admittance and impedance. *Journal of Engineering for Industry-Transactions of the ASME*, 86:9–21, 1964.

[26] M. Lievens, C. Höller, P. Dietrich, and M. Vorländer. Predicting the interaction between structure-borne sound sources and receiver structures from independently measured quantities: Case study of a washing machine on a wooden joist floor. *Acta Acustica united with Acustica*, to be published.

[27] M. Lievens and M. Vorländer. Investigation into the importance of the degrees of freedom for the characterisation of structure-borne sound sources: Case study of a washing machine on a wooden floor. *Acta Acustica united with Acustica*, 97(6):940–948, 2011.

[28] G. J. O'Hara. Mechanical impedance and mobility concepts. *J. Acoust. Soc. Am.*, 41(5):1180–1184, May 1967.

[29] S. Madhu. *Linear Circuit Analysis.* 1988.

[30] A. Moorhouse, A. Elliott, and T. Evans. In situ measurement of the blocked force of structure-borne sound sources. *Journal of Sound and Vibration*, 325(4-5):679–685, September 2009.

[31] M. Lievens. Investigation into the importance of the degrees of freedom for the characterisation of structure-borne sound sources. *Acta Acustica united with Acustica*, 96:899–904, 2010.

[32] C. Kling. *Investigations into Damping in Building Acoustics by Use of Downscaled Models.* PhD thesis, Institute of Technical Acoustics, RWTH Aachen University, 2008.

[33] A. Moorhouse. Simplified calculation of structure-borne sound from an active machine component on a supporting substructure. *Journal of Sound and Vibration*, 302(1-2):67–87, April 2007.

[34] ITA-Toolbox. www.ita-toolbox.org.

[35] ISO 9611 Acoustics - Characterization of sources of structure-borne sound with respect to sound radiation from connected structures - Measurement of velocity at the contact points of machinery when resiliently mounted, 1996.

[36] ISO 7626 Vibration and shock - Experimental determination of mechanical mobility, 1990.

[37] C. Höller. Characterization of structure-borne sound sources in buildings. Master's thesis, Institut für Technische Akustik, RWTH Aachen University, 2010.

[38] D. Ewins and J. Griffin. A state-of-the-art assessment of mobility measurement techniques–results for the mid-range structures (30-3000 hz). *Journal of Sound and Vibration*, 78(2):197–222, September 1981.

[39] M. Blau. Indirect measurement of multiple excitation force spectra by frf matrix inversion: Influence of errors in statistical estimates of frfs and response spectra. *Acta Acustica united with Acustica*, 85:464–479, 1999.

[40] A. Moorhouse, T. Evans, and A. Elliott. Some relationships for coupled structures and their application to measurement of structural dynamic properties in situ. *Mechanical Systems and Signal Processing*, 25(5):1574–1584, July 2011.

[41] T. Pritz. Transfer function method for investigating the complex modulus of acoustic materials: Spring-like specimen. *Journal of Sound and Vibration*, 72(3):317–341, October 1980.

[42] J. D. Dickens. *Dynamic Characterisation of Vibration Isolators*. PhD thesis, School of Aerospace and Mechanical Engineering, The University of New South Wales, 1998.

[43] S. Helderweirt, H. V. der Auweraer, and P. Mas. Application of accelerometer-based rotational degree of freedom measurements for engine subframe modelling. *XIX International Modal Analysis Conference*, 2001.

Curriculum Vitæ

Personal Data

11.03.1977	Matthias Lievens, born in Gent, Belgium

Education

1989–1995	Secondary Education ASO, St-Gerolfinstituut Aalter, Belgium
1995–1996	Media Technology, RITS Brussels, Belgium
1996–1997	Electronics, KAHO Gent, Belgium
1997–1998	Audio Design, Kunst Hogeschool Utrecht Hilversum, The Netherlands
1998–2001	Bachelor of Engineering in Electroacoustics, University of Salford, UK
2001–2003	Master of Engineering Sciences in Sound and Vibration, Chalmers University, Sweden
2005–2011	Ph.D. at the Institute of Technical Acoustics, RWTH Aachen University, Germany

Employments

08/2003–02/2005	Research Assistant at the Department of Phoniatrics, Pedaudiology and Communication Disorders, University Hospital Aachen, Germany
04/2005–03/2011	Research Assistant at the Institute of Technical Acoustics, RWTH Aachen University, Germany
since 09/2011	Project Engineer NVH, HEAD Acoustics GmbH, Herzogenrath, Germany

Bisher erschienene Bände der Reihe

Aachener Beiträge zur Technischen Akustik

ISSN 1866-3052

1	Malte Kob	Physical Modeling of the Singing Voice ISBN 978-3-89722-997-6 40.50 EUR
2	Martin Klemenz	Die Geräuschqualität bei der Anfahrt elektrischer Schienenfahrzeuge ISBN 978-3-8325-0852-4 40.50 EUR
3	Rainer Thaden	Auralisation in Building Acoustics ISBN 978-3-8325-0895-1 40.50 EUR
4	Michael Makarski	Tools for the Professional Development of Horn Loudspeakers ISBN 978-3-8325-1280-4 40.50 EUR
5	Janina Fels	From Children to Adults: How Binaural Cues and Ear Canal Impedances Grow ISBN 978-3-8325-1855-4 40.50 EUR
6	Tobias Lentz	Binaural Technology for Virtual Reality ISBN 978-3-8325-1935-3 40.50 EUR
7	Christoph Kling	Investigations into damping in building acoustics by use of downscaled models ISBN 978-3-8325-1985-8 37.00 EUR
8	Joao Henrique Diniz Guimaraes	Modelling the dynamic interactions of rolling bearings ISBN 978-3-8325-2010-6 36.50 EUR
9	Andreas Franck	Finite-Elemente-Methoden, Lösungsalgorithmen und Werkzeuge für die akustische Simulationstechnik ISBN 978-3-8325-2313-8 35.50 EUR

10	Sebastian Fingerhuth	Tonalness and consonance of technical sounds ISBN 978-3-8325-2536-1 42.00 EUR
11	Dirk Schröder	Physically Based Real-Time Auralization of Interactive Virtual Environments ISBN 978-3-8325-2458-6 35.00 EUR
12	Marc Aretz	Combined Wave And Ray Based Room Acoustic Simulations Of Small Rooms ISBN 978-3-8325-3242-0 37.00 EUR
13	Bruno Sanches Masiero	Individualized Binaural Technology. Measurement, Equalization and Subjective Evaluation ISBN 978-3-8325-3274-1 36.50 EUR
14	Roman Scharrer	Acoustic Field Analysis in Small Microphone Arrays ISBN 978-3-8325-3453-0 35.00 EUR
15	Matthias Lievens	Structure-borne Sound Sources in Buildings ISBN 978-3-8325-3464-6 33.00 EUR